T0094368

A Concise Introduction to

Hypercomplex Fractals

A Concise Introduction to

Hypercomplex Fractals

Fractals

Andrzej Katunin

CRC Press
Taylor & Francis Group
Boca Raton London New York

CRC Press is an imprint of the
Taylor & Francis Group, an **informa** business

CRC Press
Taylor & Francis Group
6000 Broken Sound Parkway NW, Suite 300
Boca Raton, FL 33487-2742

© 2017 by Taylor & Francis Group, LLC
CRC Press is an imprint of Taylor & Francis Group, an Informa business

No claim to original U.S. Government works

Printed in Canada on acid-free paper
Version Date: 20170315

International Standard Book Number-13: 978-1-138-63342-1 (Hardback)

Visit the Taylor & Francis Web site at
http://www.taylorandfrancis.com

and the CRC Press Web site at
http://www.crcpress.com

Contents

Preface

Since I started working with fractals, my family, friends, colleagues, and students were puzzled hearing about hypercomplex fractals. This is mainly due to the specific nomenclature used both in the fractal theory as well as in the hypercomplex number theory, with all of their attractors, repellers, octonions, zero divisors, and nilpotents. These words naturally scare them! And at the same time, they love colorful images generated by my computer simulations. These images, however, are just nice pictures to them. I am convinced that many non scientists, who had contact with complex and hypercomplex fractals have the same feelings. Therefore, I decided to write a book which would present complex and hypercomplex fractals in a concise and comprehensible manner omitting mathematical formalism as much as possible. This idea has germinated in my mind for a few years, and finally I can place the fruit of a few years' work into your hands.

This book concisely presents the complete story on complex and hypercomplex fractals, starting from the very first steps in complex dynamics and resulting complex fractal sets, through the generalizations of Julia and Mandelbrot sets on a complex plane and the Holy Grail of fractal geometry – a 3-D Mandelbrot set, and ending with hypercomplex, multicomplex, and multihypercomplex fractal sets that are still under consideration of scientists. I tried to write this book in a simple way in order to make it understandable to most people whose math knowledge covers the fundamentals of complex numbers only. Moreover, the book is full of illustrations of generated fractals and stories about great mathematicians, number spaces, and related fractals. In most cases, only information required for proper understanding of a nature of a given vector space or a construction of a given fractal set is provided; nevertheless, a more advanced reader may treat this book as a fundamental compendium on hypercomplex fractals, with references to purely scientific issues like dynamics and stability of hypercomplex systems.

The preparation of this book would not be possible without the reviewers: Professor Wojciech Chojnacki from the University of Adelaide,

Australia, who is an outstanding specialist in computer science, computer graphics, and mathematics related to these disciplines, and Dr. Krzysztof Gdawiec from the University of Silesia, Poland, who is the eminent fractal researcher working on developing new types of fractal sets. They both contributed many corrections and additions to make this book even better, and I am very grateful to them for their valuable comments and discussion.

I would like to thank my dear fiancée, Angelika, who supported me during my writing of this book, read its draft version, and provided a great feedback. I am very grateful to her for this help.

Finally, I would like to thank CRC editorial staff, especially Rick Adams, Jessica Vega, and Robin Lloyd-Starkes, for their great and professional support during the entirety of the publishing process.

Andrzej Katunin
January, 2017

CHAPTER 1

Introduction to Fractals on a Complex Plane

Fractal. This mysterious word penetrates modern pop-culture and society. But what does it mean? What is the definition of a fractal?

Actually, considering a great variety of types, shapes, and properties of objects that are called fractals, it is really hard to formulate a universal definition. In 1975, the word "fractal" was introduced by Benoit B. Mandelbrot, the father of a fractal geometry, and popularized in his famous book, *The Fractal Geometry of Nature* [79]. It comes from the Latin *fractus*, which means "broken" or "fractured." This name explains its nature, which is usually characterized by a very complex shape and, in most cases, fractional dimension. Beyond this property, fractals have few other differences from other geometrical objects. The next property is the self-similarity of fractals, which means that they looks exactly the same no matter how big the magnification of the fractals are. Roughly speaking, fractals are constructed from smaller copies of themselves. And the last thing that characterizes a fractal (which results from already presented properties) – it cannot be represented by a closed form expression, but by a recurrent dependency.

However, fractals and hierarchical structure of objects were known a long time before the Mandelbrot. The best examples are the Indian temples built in the Middle Ages[1] (see example in Figure 1.1). Many proofs of self-similarity of Hindu temples were given by numerous

---

[1]Striking examples of Indian temples that use self-similarity in their constructions are the Kandariya Mahadeva Temple built in 1030 in Khajuraho, the towers of Meenakshi Amman Temple built in 1623–1655 in Madurai, the Shveta Varahaswamy Temple built in 1673–1704 in Mysore, and many others.

Figure 1.1: A view of Chennakesava Temple, Somanathapura, India, built in 1268 (photo courtesy of Arlan Zwegers).

researchers [31, 106, 117]. Additional proofs were provided by Ron Eglash, who works in the area of ethnomathematics, and are outlined in his book [36]. He wrote that Africans have been widely using fractals in their culture (e.g., in architecture and textile design) for centuries. In the 20th century, fractals conquered arts, they appeared in works of artists who used the decalcomania technique; several paintings of the 20th-century surrealists also consist of a fractal hierarchy (e.g. *The Face of War* painted in 1940 by Salvador Dalí).

The first objects that we call now fractals appeared at the turn of the 19th and 20th centuries, and started from the simplest fractal — the Cantor set, named after Georg Cantor, a German mathematician. This fractal was constructed by Cantor in 1883, and inspired by earlier studies of Karl Weierstrass, a German mathematician who introduced everywhere continuous but nowhere differentiable function (which is known now as the Weierstrass function), the prototype of a fractal. Two decades later, the next fractal function appeared — the Koch curve, and then Koch snowflake, which was proposed by Swedish mathematician Helge von Koch in 1904.

In the meantime, two fractal curves of a special type appeared, the so-called space-filling curves. One of them was constructed by Giuseppe Peano in 1890, and the other one by David Hibert in 1891. After introducing these two curves, shortly thereafter many other self-similar space-filling curves appeared in the near time. One can get more information on these curves from [4].

The next step in fractals theory was made by Wacław Sierpiński, Polish mathematician, representative of the Warsaw school of mathematics, when he proposed his famous triangle in 1915, named after him. A year after, Sierpiński presented his carpet, which was originally discovered by his PhD student, Stefan Mazurkiewicz, in 1913. A few years later, a great milestone in fractals came through the work of two French mathematicians, Gaston Julia and Pierre Fatou. They performed investigations on properties of dynamical systems and fractal behavior associated with mapping complex numbers. This was the first step to appearance fractals on a complex plane with much greater geometrical complexity than the above presented deterministic fractals.

1.1 BIRTH OF COMPLEX FRACTALS – JULIA AND FATOU SETS

This section introduces the fundamental concepts of recursive functions that were used by Julia and Fatou to generate fractals on the complex plane as well as their properties and the simplest generalizations.

1.1.1 Preliminaries

The studies of Julia and Fatou on complex dynamic systems during the first decades of the 20th century resulted in sets with truly unique properties, which were defined from a simple recursive function, which can be written as:

$$z_{n+1} = z_n^2 + c, \tag{1.1}$$

where $z_n \in \mathbb{C}$ is an iterated element, while $c \in \mathbb{C}$ is a constant parameter. Alternatively, the form (1.1) can be presented as:

$$z \to z^2 + c \text{ or } f(z) = z^2 + c.$$

Here, we need to introduce the complex numbers $\mathbb{C}$, which can be expressed in the form of $a + bi$, where $a, b \in \mathbb{R}$, and i is an imaginary unit, which satisfies the equation $i^2 = -1$.

The field $\mathbb{C}$ of complex numbers is a 2-algebra with basis $e_0 = 1$, $e_1 = i$, which holds three important conditions (the importance of these conditions will be highlighted in the next chapters, when the hypercomplex numbers will be discussed in detail):

- commutativity

$$xy = yx, \tag{1.2}$$

- associativity

$$(xy)z = x(yz), \text{ and} \tag{1.3}$$

- alternativity

$$(yx)x = y(xx), \tag{1.4}$$

where $x, y, z \in \mathbb{C}$ (the books [98, 110] are recommended for further reading).

The behavior of the system described by (1.1) during the iteration process is like a competition between two forces. If we map the successively generated points on the complex plane (or simply a $\mathbb{C}$-plane formed by real and imaginary parts of complex numbers), we will see this battle from the top. The battlefield is divided into regions of various sizes and importance for the competing forces. In some places, a border between their territories is strait and smooth, while in other ones, a shape of a border is very complex, and each force wants to seize at least the smallest part of a territory. This border between territories of the mentioned forces is a fractal, indeed.

In the case of dynamical complex systems, these forces are the sets of points — some of them tend to infinity (escapees), while the other group of points is located in some restricted area (prisoners). The boundary between these two groups is a Julia set $\mathcal{J}$ or Fatou set $\mathcal{F}$, depending on their properties. This depends on the value of the initial point z_0 in (1.1), namely, if the generated sequence $z_0, z_1, z_2, \ldots, z_n = f(z_{n-1})$ behaves periodically or constantly, the point $z_0 \in \mathcal{F}(f)$, where $\mathcal{F}(f)$ denotes the set of Fatou functions, while if the mentioned sequence reveals irregular and highly nonlinear oscillations, in other words, behaves chaotically, then the point $z_0 \in \mathcal{J}(f)$, where $\mathcal{J}(f)$ denotes the set of Julia functions. Such sequences for every z_0 are called the orbits. If such an orbit is attracting, then it is contained in $\mathcal{F}$; while it is repelling, then it is contained in $\mathcal{J}$.

Considering that we must deal with the escapees and the prisoners, the first group must escape somewhere, while the second one should occupy some area on the $\mathbb{C}$-plane. In the case of the escapees, the successive values in the iterated sequence, starting from z_0, tend to infinity. It is convenient to present the location of these points on the

Riemann sphere, which is the representation of a $\mathbb{C}$-plane in spherical coordinates. In this case, the infinity can be represented by a single point — the north pole of the Riemann sphere. The prisoners represent periodic points, which means that there exists such a transform that $f^m(z_0) = z_0$ (note that $f^m(\cdot)$ denotes a function composition performed m times), i.e., after m iterations, the sequence reaches exactly the same point as an initial point z_0. Here, the period of such a periodic orbit is defined as the smallest value of m having the property $f^m(z_0) = z_0$. The special case is a fixed point that appears when $m = 1$. In fractals theory, the existence of two subsets of escapees and prisoners is called a *structural dichotomy*. This is also related to properties of $\mathcal{F}$- and $\mathcal{J}$-sets, i.e., the first one is a totally disconnected set (which can be considered as a generalized Cantor dust [98]), while the second one is a connected set, i.e., the $\mathcal{J}$-set encircles the subset of the prisoners. Moreover, in the case of $\mathcal{J}$-sets, there exist the so-called *fixed points* associated with c values in (1.1). These points define *attractors* of the considered complex dynamical system, i.e., the points that attract other points and form the subset of the prisoners. The area occupied by the prisoners is called the *basin of attraction*. However, there is a second basin of attraction for the escapees. Recalling the previously mentioned Riemann sphere, it is easy to see that the fixed point for the subset of escapees is the point at infinity — the north pole of the Riemann sphere, which attracts the escapees.

Special attention should be paid to prisoner points with $m = 1$. These points are the fixed points, which means that no matter how many times the iteration procedure $f^m(\cdot)$ is applied to z_0, the resulting point is always z_0. These points form $\mathcal{J}$-set that are invariant with respect to the transformation $f^m(\cdot)$. The resulting curve of a boundary between prisoners and escapees (which is the $\mathcal{J}$-set) is infinitely complex and self-similar, which means that $\mathcal{J}$ is a fractal, except for two cases: when $c = 0$ (in this case, (1.1) reduces to $z_{n+1} = z_n^2$ and the boundary has a circular shape), and when $c = -2$ (in this case, $\mathcal{J}$-set is a straight line). In order to show the existence of self-similarity in these sets, an example with several magnifications of various regions of a $\mathbb{C}$-plane was presented in Figure 1.2. The pictures represent successive magnifications, and the red frame depicts the region of the fragment of a $\mathbb{C}$-plane presented in the next picture. One can observe that the structure and degree of complexity remain the same under various magnifications. Several other examples of various $\mathcal{J}$-sets of the family defined by (1.1) are presented in Figure 1.3. In this case, the black areas are the prisoners, while the white areas are the escapees. Many of $\mathcal{J}$-sets achieved their own names, e.g., the sets presented in Figures 1.3a–d are called *San Marco fractal, Siegel Disk, Dendrite,* and *Douady rabbit,* respectively. A reader interested in the history of these sets and their importance for development of the theory of complex dynamical systems is referred to [3, 11]. Some of the above-described cases have interesting properties, e.g., Dendrite (Figure 1.3c) is a filled $\mathcal{J}$-set without interior (the prisoner points do not exist for this case). Note that the filled $\mathcal{J}$-set is a set of points z obtained during the iteration of a recursive mapping function that does not escape to infinity; this set is identical to the set of prisoners described above. In observing $\mathcal{J}$-sets, one can see that sets with axial symmetry reflect an absence of the imaginary unit in c, while the rotational symmetry is characteristic for $\mathcal{J}$-sets containing the imaginary unit in c. Moreover, depending on the sign of the imaginary unit, "swirls" of $\mathcal{J}$-sets are oriented in different directions, i.e., for negative i "swirls" are oriented clockwise (see Figures 1.3b, e), while for positive i they are oriented counter-clockwise (see Figures 1.3c, d, f).

The points that belong to a $\mathcal{J}$-set are the *neutral points* or *neutral orbits*. The dynamics of the neighborhood of neutral periodic orbits is more complicated than the dynamics of attracting or repelling orbits. In fact, they attract the points in the local neighborhood, and depending on the number of periods the characteristic trajectories (in the form of petals of flowers) appear. Their behavior is described by the so-called *Flower Theorem* (see [11] for details).

The $\mathcal{J}$-sets presented in Figure 1.3 are black and white; however, most readers familiar with fractals on a complex plane, or those who at least have seen such fractals before, remember these sets as images full of beautiful colors. A color representation is also very useful in some cases, e.g., when filled $\mathcal{J}$-sets can be easily presented in black and white, the $\mathcal{F}$-sets need to be presented with some surroundings, otherwise we will obtain just a set of disconnected points on a $\mathbb{C}$-plane. Therefore, it is necessary to introduce here how to obtain the colorful $\mathbb{C}$-plane with $\mathcal{J}$- and $\mathcal{F}$-sets.

1.1.2 Coloring Fractal Sets on a Complex Plane

In order to obtain colorful representations of $\mathcal{J}$- and $\mathcal{F}$-sets several algorithms have been developed to date. The most popular one is the *escape-time algorithm* widely used in fractal generating and rendering software. The escape-time algorithm discretely colors pixels of a $\mathbb{C}$-plane, and the colors denote how quickly a given point escapes to infinity. The

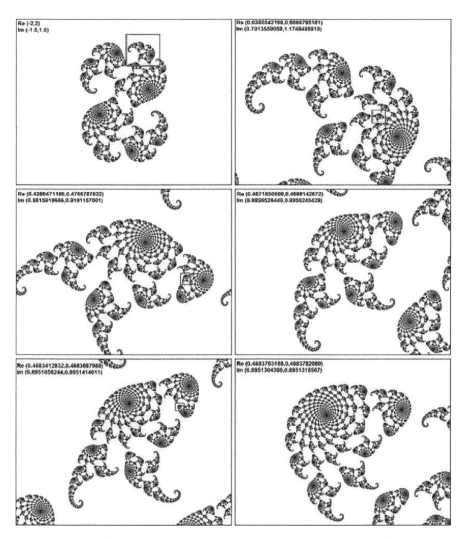

Figure 1.2: $\mathcal{J}$-set with $c = 0.3 + 0.022i$ under various magnifications.

black color is usually used for the representation of prisoner points (i.e., the interior of a $\mathcal{J}$-set). The algorithm works as follows. Starting from the initial point z_0, a sequence is generated, and at every iteration, the conditions for this sequence are checked. The first condition is a criticality of achieved values called *bailout condition*. When the critical value is reached, the algorithm stops, and a pixel is assigned with a defined color, and the next point is checked. Since $|z_0| > 2$ and $2 \notin \mathcal{J}$, usually the bailout value is assumed to be equal 2. In other words, the color of a given pixel is

(a) $c = -0.75$

(b) $c = -0.390541 - 0.586788i$

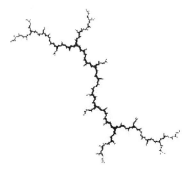

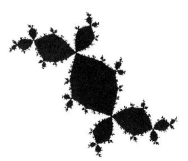

(c) $c = i$

(d) $c = -0.123 + 0.745i$

(e) $c = 0.27334 - 0.00742i$

(f) $c = -1 + 0.1i$

Figure 1.3: Examples of $\mathcal{J}$-sets for various values of c.

related to the number of iterations that allow escaping of a given point to infinity. Obviously, during coloring of such sets using computer software, the number of iterations should be finite. However, the selected number of iterations is related to the accuracy of the obtained approximation of a fractal boundary, i.e., the higher number of iterations the more accurate the boundary is, but the more computational time is required.

Although the escape-time algorithm is very simple and effective, the fact that the number of iterations is always an integer value causes that the aliasing effect may appear in the escaping area, i.e., this area is divided into bands of constant colors (see Figure 1.4a). For the presented $\mathcal{J}$-set in Figure 1.4, the parameter $c = 0.29 + 0.01i$. In order to avoid this effect and enable coloring a $\mathbb{C}$-plane in a continuous way, several techniques have been developed. Most of them are based on the *distance estimators* concept. Barrallo and Jones [7] mentioned three types of distance estimators: *classical distance estimation algorithm*, *continuous potential algorithm*, and *normalized iteration count algorithm*. All of them are based on continuous interpolation of colors in bands. The results of such interpolation are presented in Figure 1.4b.

A group of *escape-angle algorithms* (or *binary decomposition algorithms*) was introduced in order to represent not only the velocity, but also the angles of escape. The example of using this algorithm with quaternary decomposition (the black and white colors are assigned alternately to each quadrant of a $\mathbb{C}$-plane) is presented in Figure 1.4c. Similarly to the escape-time algorithm, the escape-angle algorithm can be transformed into a continuous decomposition algorithm by applying distance estimators or other interpolation techniques.

The other coloring techniques include statistical transformations of sequences $z_0, z_1, z_2, \ldots, z_n = f(z_{n-1})$, i.e., the statistical functions like mean, median, standard deviation, etc., are applied to the obtained magnitudes of these sequences, and the resulting values are directly applied for coloring. This approach is often available in the fractal generating and rendering software. Another popular family of coloring algorithms is *orbit traps*. The simplest version of the algorithm from this family is based on selection of a region in a $\mathbb{C}$-plane and observation of the relationship between z_n values and values from this region. Such a region is usually defined as a simple shape and a threshold distance. If z_n falls inside the trap, the iterating procedure stops and the pixel is colored based on the determined distance to the trap. The examples of the application of orbit trap algorithms can be found in [14, 121].

In addition, Barrallo and Jones [7] mentioned coloring of fractals

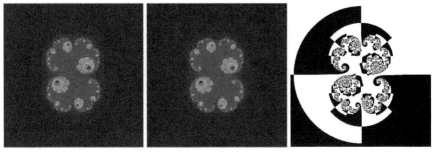

(a) escape-time algorithm   (b)   distance   estimation   (c) escape-angle algorithm
algorithm

Figure 1.4: Coloring of fractal sets on a complex plane using various techniques.

using the *finite attractor algorithm*. Recalling the fixed points discussed in section 1.1.1, one can remember that the fractal sets contain such points that attract the other and create the basin of attraction. This property is used for coloring both the interiors and exteriors of fractal sets on a $\mathbb{C}$-plane. The value used for assigning a color for a given point is the convergence rate to the nearest fixed point. More advanced techniques (already used in many fractal generating and rendering software) consider combining the above-presented ones. Following the nomenclature introduced in [8], such algorithms are called the *multilayer coloring algorithms*. The examples of using these algorithms can be found in the above-cited literature.

Using the most common escape-time algorithm, some visually attractive $\mathcal{F}$- and $\mathcal{J}$-sets described by (1.1) are visualized in Figure 1.5 with various color palettes.

Additionally, the algorithms for coloring the interiors of filled $\mathcal{J}$-sets have been developed. The color palette is applied based on velocity of attracting to critical points (the points where the function's derivatives vanish), and most of the above-mentioned algorithms are applicable for this purpose. Another way of coloring the interior of filled $\mathcal{J}$-sets is based on the number of periods of orbits. In this case, the interior of each single area (like in Douady rabbit, presented in Figure 1.3d) is filled by one color.

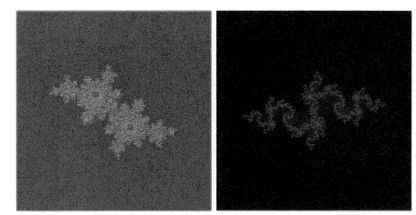

(a) $c = -0.4 + 0.6i$ (b) $c = -0.835 - 0.2321i$

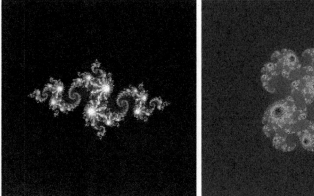

(c) $c = -0.8 + 0.156i$ (d) $c = 0.3 + 0.022i$

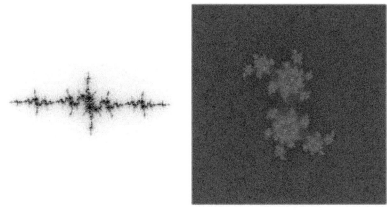

(e) $c = -1.3 + 0.06i$ (f) $c = 0.35 + 0.45i$

Figure 1.5: Examples of colored $\mathcal{F}$- and $\mathcal{J}$-sets for various values of c.

1.1.3 Julia and Fatou Sets within the Polynomials of Various Orders

Previously, we considered the complex quadratic polynomial given by (1.1), which is the most popular form of polynomial maps of fractal sets on a $\mathbb{C}$-plane, but it is only a special case of a general form

$$z \rightarrow z^p + c \text{ for } p \geq 2, \tag{1.5}$$

where p denotes the power of z. Taking into consideration such a generalization allows for sufficiently enlarging the number of possible $\mathcal{F}$- and $\mathcal{J}$-sets. Several studies of these functions were performed and described in [47, 48].

The generalized form of $\mathcal{J}$- and $\mathcal{F}$-sets given by (1.5) has many interesting properties. The point at infinity for (1.5) is a super-attracting fixed point (its derivative is equal 0 and it belongs to $\mathcal{F}$-set — see [76] for details), and considering that for every $z \neq \infty$ Equation (1.5) has a finite value, this fixed point is also an exceptional point. Moreover, in open $\mathbb{C}$-plane there exists only one singular point and at most $p-1$ critical fixed points. Thus, at most $p-1$ stable orbits (the orbits that remain almost invariant around a fixed point independently of a value of an iterated function) [70]. The discussion with proofs of the above statements can be found in [11].

The selected $\mathcal{J}$-sets for various orders of an iterated polynomial are presented in Figure 1.6. For clarity, the symbols describing these sets are introduced, e.g., $\mathcal{J}_3(-0.19, 0.86)$ denotes a $\mathcal{J}$-set with $p = 3$, and $c = -0.19 + 0.86i$. The order of rotational symmetry for $\mathcal{J}$- and $\mathcal{F}$-sets with imaginary unit in c is equivalent to p, i.e., the sets with an exponent p have $p-1$-order rotational symmetry with respect to their origin; while for sets with a real part of c, only the axial symmetry is preserved. Moreover, the directionality of "swirls" remains the same as in the quadratic family (see Section 1.1.1).

Considering the increased order of symmetry of a $\mathcal{J}$-set with an increase of p, the natural question is what a $\mathbb{C}$-plane will look like when $p \rightarrow \infty$. When analysing $\mathcal{J}$-sets of polynomials of lower-order (see Figures 1.7a–d), it is difficult to deduce what a shape of $\mathcal{J}$-set it is for higher values of p; however, when looking at $\mathcal{J}$-set with higher value of p (see Figures 1.7e–h), one can observe that a shape of a $\mathcal{J}$-set tends to be a circle. In fact, for very high values of p, a $\mathcal{J}$-set looks like a circle more and more. However, if p is finite, the resulting boundary is still a fractal. But if $p \rightarrow \infty$, then the resulting boundary is a circle, and, obviously, stops being a fractal. What is also interesting, when $p \rightarrow \infty$, the resulting boundary is always circular, regardless of c (see [13] for details). This phenomenon has quite a simple explanation. For $p \to \infty$, the resulting successive values of z_n rapidly increase with the increase of n when $|z_n| > 1$, and rapidly decrease with the decrease of n when $|z_n| < 1$. The influence of the parameter c becomes insignificant when considering sequences z_n. Following this, when $p \to \infty$, the resulting boundary is a unit circle. Despite this, the behavior of orbits that belong to this set remains chaotic.

The next, quite natural, question is why (1.5) is limited by the condition for p. Let us consider three cases for p value: (1) when $0 \leq p < 1$, (2) when $p = 1$, and (3) when $1 < p < 2$. In the first case, the critical point is no longer located in zero (for sets described by (1.5), we have a critical point in zero), the critical point is located at infinity. The parameter c is not on the trajectory of ∞, therefore, the $\mathcal{J}$-set does not exist for $0 \leq p < 1$. In the second case, the critical point does not exist, so the $\mathcal{J}$-set also does not exist [118]. In the third case, the critical point comes back to zero, the resulting structures resemble the integer-valued $\mathcal{J}$- and $\mathcal{F}$-sets and hold the axial symmetry (along the axis of reals), while the rotational symmetry is broken. The resulting sets, however, hold their fractal properties for arbitrary $p > 1$, $p \in \mathbb{Q}$ ($\mathbb{Q}$ denotes the rational numbers), which was proven by Liu et al. [77]. The pioneering studies on fractional p were performed by Gujar et al. [34, 48], where they explored the structure of fractal sets on a complex plane for $p \in \mathbb{R}$. In the interval $1 < p < 2$, there is a family of graphically interesting fractal sets − the Glynn sets, named after E. F. Glynn [42, 102]. This family is based on $p = 1.5$, which creates tree-like fractal sets (see Figure 1.8).

And finally, let us consider the case when p is negative. This case was studied by Shirriff [115], who was probably the first investigator of $\mathcal{J}$-sets for negative p. Considering the nonexistence of $\mathcal{J}$-sets for $-1 < p < 0$ (similarly to $0 \leq p < 1$), and broken symmetry for $-2 < p < -1$, let us consider the case when $p < -2$, $p \in \mathbb{Q}$. Surprisingly, the dynamics of complex systems for $p < -2$ is completely different with respect to positive exponents of an iterated variable. In analyzing such systems, one can observe that the critical point of such a set is at infinity, while there is no critical point at zero, as in systems described by (1.5), and it is true for all $p < -1$, $p \in \mathbb{Q}$. Recalling the Riemann sphere, the resulting set looks like it was turned inside out. This means that the set of prisoners has an attractor at the infinity point on the Riemann sphere, and its basin of attraction covers the whole Riemann sphere except some small regions near c. Therefore, the obtained sets are not $\mathcal{J}$-sets by the definition (the

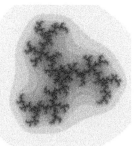

(a) $\mathcal{J}_3\,(-0.19, 0.86)$ (b) $\mathcal{J}_3\,(-0.12, 0.8)$ (c) $\mathcal{J}_3\,(-0.63, 0.29)$

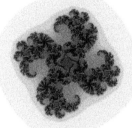

(d) $\mathcal{J}_4\,(-0.6, 0.4)$ (e) $\mathcal{J}_4\,(-0.3, 0.46)$ (f) $\mathcal{J}_5\,(-0.74, 0.81)$

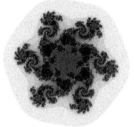
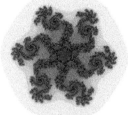

(g) $\mathcal{J}_6\,(-0.5, 0.58)$ (h) $\mathcal{J}_6\,(-0.7, 0.3)$ (i) $\mathcal{J}_8\,(0, 0.82)$

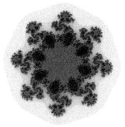

(j) $\mathcal{J}_8\,(-0.65, 0.52)$ (k) $\mathcal{J}_{12}\,(-0.69, 0.52)$ (l) $\mathcal{J}_{20}\,(-0.46, 0.8)$

Figure 1.6: Examples of $\mathcal{J}$-sets of various polynomial orders.

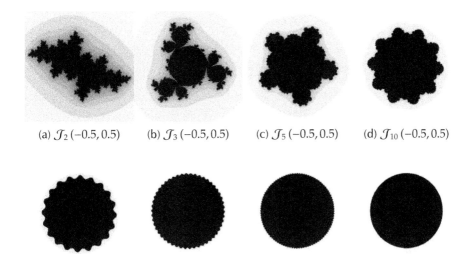

(a) $\mathcal{J}_2(-0.5, 0.5)$ (b) $\mathcal{J}_3(-0.5, 0.5)$ (c) $\mathcal{J}_5(-0.5, 0.5)$ (d) $\mathcal{J}_{10}(-0.5, 0.5)$

(e) $\mathcal{J}_{20}(-0.5, 0.5)$ (f) $\mathcal{J}_{50}(-0.5, 0.5)$ (g) $\mathcal{J}_{100}(-0.5, 0.5)$ (h) $\mathcal{J}_{1000}(-0.5, 0.5)$

Figure 1.7: Tending of $\mathcal{J}_p(-0.5, 0.5)$ to a circular-like shape.

$\mathcal{J}$-sets must have two critical points in zero and infinity). Such sets will be further called the *degenerated $\mathcal{J}$-sets*. However, these sets hold several properties of positive exponent-$\mathcal{J}$-sets. The obtained sets are rotationally symmetric, i.e., the sets generated with an exponent $-p$, $p \in \mathbb{Z}$, have p-order rotational symmetry with respect to their origin. But, in general, these sets are not axially symmetric unless c is real [115].

Coloring of degenerated $\mathcal{J}$-sets with negative exponents also differs. Since the only critical point is located at infinity, the classical escape-time algorithm generates artefacts. Therefore, it is suitable to start a sequence not from $z_0 = \infty$, but from $z_0 = c$, which, in fact, yields the same result. Shirriff [115] proposed two alternative methods of coloring of such sets: based on Lyapunov exponents determined for the mapping, which reflects stability or chaotic behavior in particular regions of a $\mathbb{C}$-plane; and based on periodicity of resulting sequences, assigning the color depending on the number of cycles. The examples of such sets are presented in Figure 1.9. More information on analysis and coloring of negative exponent-$\mathcal{J}$-sets can be found in [48, 115].

In conclusion, we can say that the $\mathcal{J}$-sets or degenerated $\mathcal{J}$-sets with fractal properties can be constructed using the following formula:

$$z \to z^p + c \text{ for } p < -1 \lor p > 1, p \in \mathbb{Q}. \tag{1.6}$$

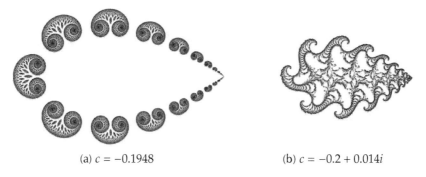

(a) $c = -0.1948$ (b) $c = -0.2 + 0.014i$

Figure 1.8: Examples of Glynn sets.

1.1.4 Other Variations of Julia and Fatou Sets on a Complex Plane

Besides the discussed classical $\mathcal{J}$- and $\mathcal{F}$-sets given by (1.1) or their generalization with respect to the exponent of polynomials given by (1.6), a lot of studies were performed in order to find other forms of polynomials and perform their mathematical and geometrical analysis. However, the main reason is to find aesthetically pleasing and visually interesting patterns. With the development of computational technologies, and the appearance of fractal generation and rendering software, many enthusiasts started working on searching for newer and newer formulas for complex dynamical systems.

A class of such systems is based on transcendental functions, i.e., the exponential, logarithmic, trigonometric, hyperbolic, and several other functions as well as inverses of all of the mentioned ones. Due to the great variety of possible modifications of (1.6) by applying various transcendental functions, only few cases will be introduced here. A reader interested in such a class of functions is referred to several papers that present analysis of dynamics of systems described by these functions [9, 37, 54, 55, 83, 111, 113] and some visualization studies [93, 105]. It was proven that sine-, cosine-, and exponential-based functions have an infinite number of critical points [33, 83]. Moreover, the point at infinity becomes an essential singularity. Several examples of $\mathcal{J}$- and $\mathcal{F}$-sets based on transcendental functions are presented in Figure 1.10.

Besides the Picard iteration scheme presented above (iterating the recursive functions in the form $z_{n+1} = f(z_n)$), several other schemes have been introduced to date. The oldest and the most known iteration scheme

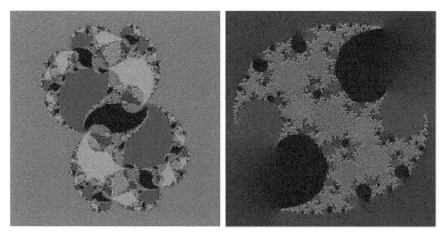

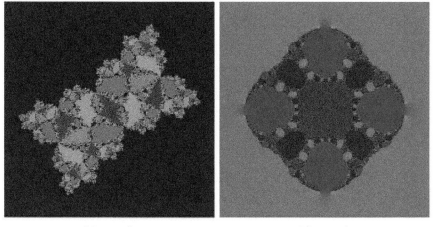

(a) $p = -2$,
$c = 0.72 + 1.05i$

(b) $p = -2$,
$c = 0.106 + 0.619i$

(c) $p = -2$,
$c = -0.73 + 0.22i$

(d) $p = -4$,
$c = -0.791$

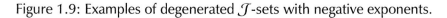

Figure 1.9: Examples of degenerated $\mathcal{J}$-sets with negative exponents.

besides the Picard iteration is the Mann iteration, named after W. Robert Mann [80]. It is given by

$$z_{n+1} = (1 - a_n) z_n + a_n f(z_n) \text{ for } n = 0, 1, 2, \ldots, \quad (1.7)$$

where $a_n \in (0, 1]$ is a certain parameter. If the sequence $a_n = \lambda$ (const), $\lambda \in [0, 1]$, the Mann iteration reduces to Kransosel'ski iteration (named after the scheme presented in [67]), and when $a_n = 1$, it reduces to the

Picard iteration [104]. The Mann iteration scheme, in the context of the generation of $\mathcal{J}$-sets, was discovered by Rani and Kumar [104], who call them "superior Julia sets."

The next iteration scheme, introduced by Ishikawa [56] and named after him, uses two parameters and is presented in the form

$$\begin{cases} z_{n+1} = (1 - a_n) z_n + a_n f(v_n), \\ v_n = (1 - b_n) z_n + b_n f(z_n) \text{ for } n = 0, 1, 2, \ldots, \end{cases} \tag{1.8}$$

where $a_n \in (0, 1]$ and $b_n \in [0, 1]$. The generation of $\mathcal{J}$-sets using this iteration scheme was described in [20]. Both Mann and Ishikawa iteration schemes were modified in several ways, which resulted in the appearance of the modified Mann and Ishikawa iterations (obtained by replacing operator $f(\cdot)$ by its m-th iterate $f^m(\cdot)$), Mann and Ishikawa iterations with errors (obtained by setting certain parameters to zero), and others. For more information on these iterations and full mathematical descriptions, the reader is referred to Berinde's book [10].

Furthermore, several other iterations were used for generation of $\mathcal{J}$-sets, namely, Noor iteration [92], Jungck Mann and Jungck Ishikawa [91], Jungck Noor [60], etc. As one can see, the structure of a recursive formula can be constructed in on almost arbitrary way, so the possibilities of creation of new fractal shapes are huge.

The last group of $\mathcal{J}$-sets worth attention is a group of biomorphs introduced by Pickover. The biomorphs are certain modifications of $\mathcal{J}$-sets (see [101] for mathematical details). Following the idea of Pickover, the biomorphs mimic simple biological systems. Biomorphs are found accidentally as a programming bug [57]. Besides their aesthetic values [75], they have several other applications: the authors of [85] used biomorphs to examine the evolution of unicellular organisms, while Levin [73] used biomorphs to describe and model embryogenesis processes. The recent studies of Gdawiec et al. [40] link the modified Pickover iterations (Mann and Ishikawa iterations) and biomorphs to generate new fractal patterns. Several examples of the new biomophs generated following these iteration schemes can be found in the latter reference.

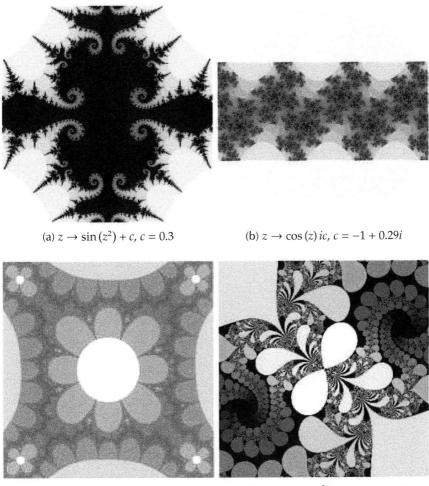

(a) $z \to \sin(z^2) + c$, $c = 0.3$

(b) $z \to \cos(z) ic$, $c = -1 + 0.29i$

(c) $z \to 1/\sinh(z^2) + c$, $c = 0.01i$

(d) $z \to \exp(c/z^2)$, $c = -0.4 + 0.7i$

Figure 1.10: Examples of $\mathcal{J}$ based on transcendental functions.

1.2 FRACTAL REVOLUTION – THE MANDELBROT SET

1.2.1 History and Concept of the Mandelbrot Set

Despite the majority of contributions Gaston Julia and Pierre Fatou made to the understanding of complex dynamical systems behavior, their studies had been ignored and forgotten by the mathematical community, until Mandelbrot started working on fractals and dynamical systems in late 1970s. From the previous section, one knows that the behavior and resulting distribution of subsets of prisoners and escapees on a $\mathbb{C}$-plane depend on c values. By changing the c value in (1.1), one can generate countably infinitely many different patterns on a $\mathbb{C}$-plane. Mandelbrot analyzed the resulting Julia and Fatou sets based on (1.1) and decided to index all possible $\mathcal{J}$- and $\mathcal{F}$-sets with respect to varying c. When he mapped the result during his experiment in 1979, the general set containing all possible Julia and Fatou sets appeared. Nowadays, this set is called after its creator – the Mandelbrot set or, simply, $\mathcal{M}$-set.

The $\mathcal{M}$-set is probably the most recognizable mathematical picture ever created by a human, and it has the most complex shape that can ever be seen [98]. It is presented as a collection of points on a $\mathbb{C}$-plane, where each point is related to $\mathcal{J}$- or $\mathcal{F}$-set for a specific c. $\mathcal{J}$- and $\mathcal{F}$-sets are closely related to the $\mathcal{M}$-set, i.e., the $\mathcal{M}$-set consists of all possible $\mathcal{J}$-sets (which are connected), while $\mathcal{F}$-sets are located outside the $\mathcal{M}$-set on a $\mathbb{C}$-plane (which are totally disconnected) – see Figure 1.11. Moreover, there are other relations between $\mathcal{J}$- and $\mathcal{M}$-sets: the shape of a $\mathcal{J}$-set is related to the shape of an $\mathcal{M}$-set corresponding with a c value of $\mathcal{J}$-set, e.g., the $\mathcal{J}$-sets with horizontally oriented elongated shapes are related to the values on the very left of a $\mathbb{C}$-plane, the circular-like shaped $\mathcal{J}$-sets are related to regions of the $\mathcal{M}$-set near the origin of a $\mathbb{C}$-plane, and the Dendrite $\mathcal{J}$-set (for $c = i$) is located on the "antenna" of one of the "bulbs." The $\mathcal{M}$-set can also be presented in the form of kind of a map that contains $\mathcal{J}$- and $\mathcal{F}$-sets depending on a discretised c value (see Figure 1.12). One can observe that $\mathcal{J}$-sets form an approximation of the shape of an $\mathcal{M}$-set, and this approximation appears when the net of $\mathcal{J}$-sets on a $\mathbb{C}$-plane is denser.

1.2.2 Geometrical Structure and Atoms of the Mandelbrot Set

From the geometrical analysis of the $\mathcal{M}$-set, one can find out that it contains a large cardioid, slightly moved to the left from the origin of a $\mathbb{C}$-plane (which ranges from -0.75 to 0.25 on the real axis), the largest bulb on the left (with a circular shape of radius 0.25 centred at $-1 + 0i$), and countably infinitely many bulbs around the perimeter of the cardioid and the largest bulb. Moreover, every bulb has its own antennas (also called filaments). When magnifying the $\mathcal{M}$-set near its antennas, one can observe the smaller copies of itself (known as secondary $\mathcal{M}$-sets). The largest secondary $\mathcal{M}$-set is located on its antenna (see Figure 1.11). The complexity of the $\mathcal{M}$-set becomes even higher than its initial form when it is magnified. Several fragments obtained from magnification of the $\mathcal{M}$-set

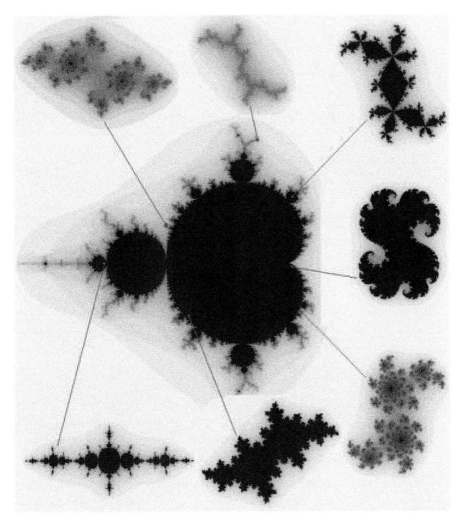

Figure 1.11: An $\mathcal{M}$-set with corresponding $\mathcal{J}$- and $\mathcal{F}$-sets on a $\mathbb{C}$-plane.

are presented in Figures 1.13–1.15. Some of them obtained their own names, like the Seahorse Valley (a portion of the $\mathcal{M}$-set centred at around $-0.75 + 0.1i$; see Figure 1.16) or the Elephant Valley (a portion of the $\mathcal{M}$-set centred at around $0.1 + 0.1i$; see Figure 1.17).

Probably the most frequently appearing elements of the $\mathcal{M}$-set are its bulbs (secondary, tertiary, etc.); which clearly point to the self-similarity of the $\mathcal{M}$-set. They are attached directly to the main cardioid and have several interesting properties. The obvious one is that the $\mathcal{M}$-set is symmetric with respect to the axis of reals, and, thus, bulbs above and below the axis

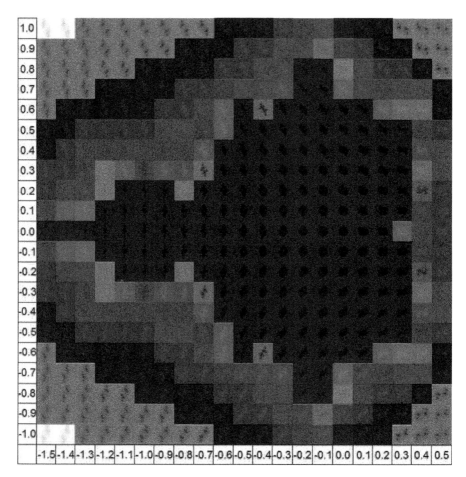

Figure 1.12: Mapping of Julia sets with varying c on a $\mathbb{C}$-plane. The values on the horizontal axis denote real parts of c, while on the vertical axis they denote imaginary parts of c.

look exactly the same. The bulbs of the $\mathcal{M}$-set can be considered as its atoms, which are connected with periods of periodical orbits generated by the sequence (1.1). The cardioid (which can be considered as a primary bulb or a primary atom) is related to a cycle of a period 1; next, the biggest bulb on the left of the cardioid is related to a cycle of a period 2; two smaller bulbs located perpendicularly to the axis of reals and attached to the cardioid are related to a cycle of a period 3, etc. Obviously, there are the cycles of higher periods attached to the cardioid and the biggest bulb (see Figure 1.18, for instance).

One can also observe that the bulbs are directed in a different

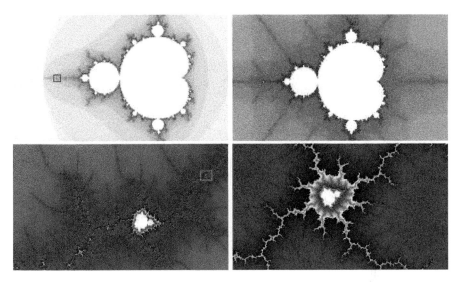

Figure 1.13: Magnifying the $\mathcal{M}$-set: secondary $\mathcal{M}$-sets.

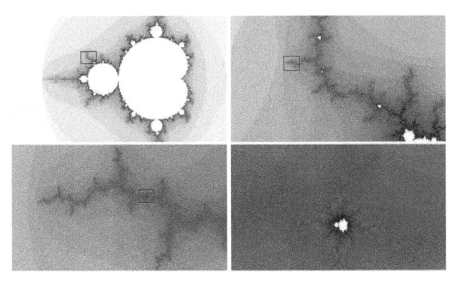

Figure 1.14: Magnifying the $\mathcal{M}$-set: filaments.

manner, and analyzing Figure 1.18, one can notice the relationship of their orientation with the periods. Indeed, the bulbs are attached to the cardioid and the largest bulb at an *internal angle* $\phi = 2\pi\,(m/n)$, where m/n is a *rotation number* that corresponds to certain angles, e.g., $1/2 \to 180°$, $1/3 \to 120°$, $1/4 \to 90°$, etc. [32]. Moreover, it is possible to obtain m_x/n_x

Figure 1.15: Magnifying the $\mathcal{M}$-set: bulbs.

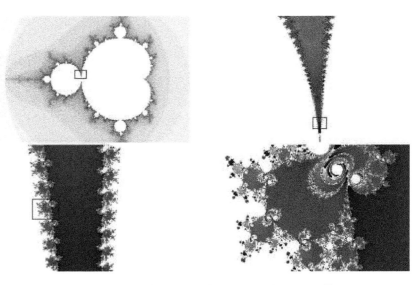

Figure 1.16: Magnifying the $\mathcal{M}$-set: Seahorse Valley.

rotation number for the bulb located between m_1/n_1 and m_2/n_2 bulbs by adding their numerators and denominators following the Farey addition rule [32], e.g., $3/8 = 1/3 + 2/5$. This is related to the Fibonacci property, e.g., the largest bulb between the bulb of a period 2 and the bulb of a period 3 has a period 5; the largest bulb between two bulbs of periods

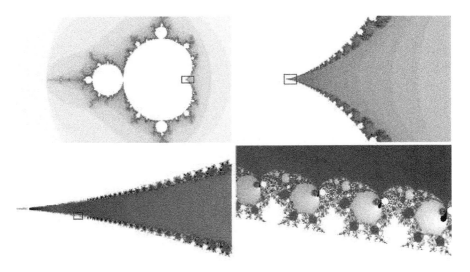

Figure 1.17: Magnifying the $\mathcal{M}$-set: Elephant Valley.

3 and 5 has a period 8, etc. This property is satisfied for secondary and higher-degree bulbs in the $\mathcal{M}$-set.

1.2.3 Mandelbrot Set within the Polynomials of Various Orders and Other Variations

Similarly, as for $\mathcal{J}$- and $\mathcal{F}$-sets, the $\mathcal{M}$-set can be described within the higher-order polynomials given by (1.5). What is interesting, is the higher-order $\mathcal{M}$-sets remain connected regardless of the order, which means that they have many common properties with higher-order $\mathcal{J}$- and $\mathcal{F}$-sets described in Section 1.1.3, as well as with a classical $\mathcal{M}$-set. Additionally, the generalized $\mathcal{M}$-sets reveal $p-1$-order rotational symmetry with respect to the origin, and some of the higher-order $\mathcal{M}$-sets are presented in Figure 1.19. Considering the case where $p \to \infty$, the $\mathcal{M}$-set tends to a circle on a $\mathbb{C}$-plane [66, 96].

One can see that the higher-order $\mathcal{M}$-sets have components corresponding to the classical $\mathcal{M}$-set, i.e., the bulbs and filaments. They also have Seahorse and Elephant Valleys. However, the rule of appearance of secondary bulbs for higher-order $\mathcal{M}$-sets slightly differs. While the secondary bulbs increase according the power series $2, 4, 8, 16, \ldots$ (i.e., $2^1, 2^2, 2^3, 2^4, \ldots$), the secondary bulbs of the higher-order $\mathcal{M}$-sets increase according the series $p^1, p^2, p^3, p^4, \ldots$, that is, correspond to the order p.

The formulation of the $\mathcal{M}$-set can be generalized with respect to

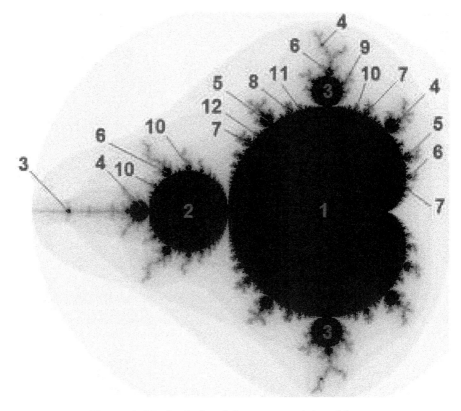

Figure 1.18: Periods of the atoms of the $\mathcal{M}$-set.

the order of a polynomial for the cases where $p \in \mathbb{Z}^-$, as well as for the cases given by (1.6). The behavior of the resulting generalized $\mathcal{M}$-sets is the same as for their analogues of the generalized $\mathcal{J}$- and $\mathcal{F}$-sets (see Section 1.1.3). Several examples of $\mathcal{M}$-sets of negative-order polynomials are presented in Figure 1.20. In contrast to positive-order $\mathcal{M}$-sets, negative-order ones have $p + 1$-order rotational symmetry with respect to the origin. However, similarly to the positive-order $\mathcal{M}$-sets, the axial symmetry along the axis of reals is preserved for all $p \in \mathbb{Z}^-$, and, additionally, the axial symmetry along the axis of imaginaries occurs for even values of p. The properties of the negative-order $\mathcal{M}$-sets ($\mathcal{M}^-$-sets), with respect to the positive-order analogues, are different, e.g., the $\mathcal{M}^-$-sets are not compact and extend out to infinity (see details in Section 1.1.3).

Among the discussed modifications of the $\mathcal{M}$-sets, there are several variations that became popular. The example of such variations is the

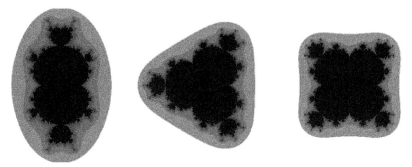

(a) $\mathcal{M}_3$ (b) $\mathcal{M}_4$ (c) $\mathcal{M}_5$

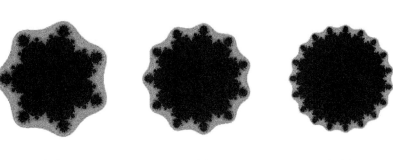

(d) $\mathcal{M}_8$ (e) $\mathcal{M}_{12}$ (f) $\mathcal{M}_{20}$

Figure 1.19: Higher-order $\mathcal{M}$-sets.

family of Mandelbar sets introduced by Crowe et al. [29], sometimes called the Tricorn $\mathcal{T}$, due to the characteristic shape of this set. The $\mathcal{T}$-set is given by

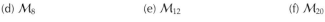

$$z \to \bar{z}^2 + c, \qquad (1.9)$$

where $\bar{z}$ is a complex conjugate of z, i.e., $\bar{z} \equiv a - bi$. The $\mathcal{T}$-set satisfies several properties of the $\mathcal{M}$-set, e.g., it is connected [89] and can be generalized with respect to the degree of antiholomorphic polynomials that describe them. Several studies of such generalizations were presented by Nakane and Schleicher [90]. Several examples of the higher-order $\mathcal{T}$-sets are shown in Figure 1.21. Some visually interesting higher-order $\mathcal{T}$-sets can also be found in [61]. Similarly to $\mathcal{M}^-$-sets, the $\mathcal{T}$-sets have $p + 1$-order rotational symmetry with respect to the origin and the same axial symmetries depending on parity of p.

    Several geometrical objects that imitate $\mathcal{M}$-set are known. One of them is a Randelbrot set which is an object obtained in the same way as

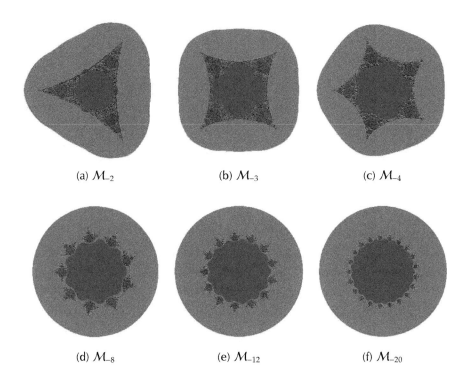

(a) $\mathcal{M}_{-2}$ (b) $\mathcal{M}_{-3}$ (c) $\mathcal{M}_{-4}$

(d) $\mathcal{M}_{-8}$ (e) $\mathcal{M}_{-12}$ (f) $\mathcal{M}_{-20}$

Figure 1.20: Examples of the $\mathcal{M}$-sets of negative-order polynomials.

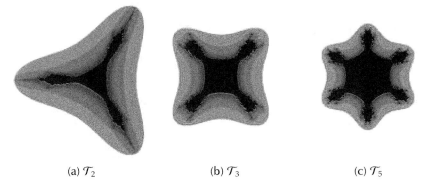

(a) $\mathcal{T}_2$ (b) $\mathcal{T}_3$ (c) $\mathcal{T}_5$

Figure 1.21: Examples of the higher-order $\mathcal{T}$-sets.

the classical $\mathcal{M}$-set, with an additional random component R added to the recursive sequence proposed by Robert M. Dickau:

$$z \rightarrow z^2 + c + R, \ R \in \mathbb{C}, \ R \equiv R_x + iR_y. \tag{1.10}$$

1.3 SEARCHING FOR THE HOLY GRAIL – 3-D MANDELBROT SET

After introducing the $\mathcal{M}$-set by Mandelbrot, numerous scientists and enthusiasts tried to find a 3-D extension of this set. However, they encountered a problem – the complex numbers cannot be directly extended to the 3-D space. This problem originates from the times of Sir W. R. Hamilton, the inventor of the algebra of quaternions. Working with problems of plane geometry and mechanics, he clarified the formulation of complex numbers, and then tried to expand the algebra of complex numbers to a 3-D space. His idea was to construct a complete number system from the triples of the real numbers. He spent a few years on this idea without success, and finally abandoned efforts on solving this problem. The main problem lies in the multiplication operation of such triples – like in the case of complex numbers this operation needed to be componentwise, but it was not possible to extend the complex numbers. After the unsuccessful studies on the complex triplets, he discovered the 4-D extension – the quaternions.

Obtaining the 3-D analogue of the $\mathcal{M}$-set was an obsession for many mathematicians [6]. Fortunately, several studies led to discovery of quasi-3-D $\mathcal{M}$-sets. The first studies were generally based on mapping the $\mathcal{M}$-set into the 3-D space using simple and well-known mapping functions. Barrallo [6] mentioned two examples: the height field based on potential of the $\mathcal{M}$-set, and projection of a $\mathbb{C}$-plane with the $\mathcal{M}$-set onto the Riemann sphere.

The breakthrough in this matter was made by a fractal enthusiast, Daniel White, in 2007. He proposed a method of implementing (1.1) in $\mathbb{R}^3$ under the assumption that the rotation appearing during squaring of a complex number (as in (1.1)) is equivalent to representation of this operation using the spherical coordinate system. Following this, triplet value in the spherical coordinate system is given by $\{\rho, \phi, \theta\}$, where ρ is a modulus, ϕ is a longitude angle, and θ is a latitude angle. The squaring operation of such value gives [6]

$$\{\rho, \phi, \theta\}^2 = \{\rho^2, 2\phi, 2\theta\}. \tag{1.11}$$

The final set of recursive equations that allow reproducing the $\mathcal{M}$-set in a

3-D space is given by

$$\begin{cases} x_{n+1} = x_n^2 - y_n^2 - z_n^2 + c_1, \\ y_{n+1} = 4x_n y_n z_n / \sqrt{y_n^2 + z_n^2} + c_2, \\ z_{n+1} = 2x_n \left(-y_n^2 + z_n^2\right) / \sqrt{y_n^2 + z_n^2} + c_3, \end{cases} \tag{1.12}$$

which gives an original 3-D $\mathcal{M}$-set presented in Figure 1.22a, and Figures 1.22b–d represent its modified versions with swapped sine and cosine functions in (1.12).

(a)   (b)

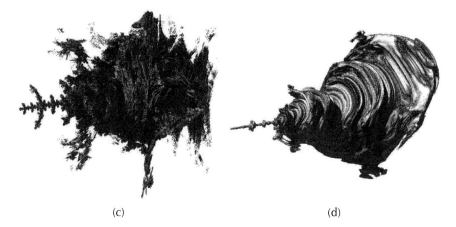

(c)   (d)

Figure 1.22: Original (a) and modified (b–d) versions of the 3-D $\mathcal{M}$-set.

One should mention the second contributor to the 3-D $\mathcal{M}$-set, Paul Nylander, also a fractal enthusiast, who improved a formulation made by

White. This is the reason why (1.12) is often called the White–Nylander formula. The obtained object is not a direct extension of the $\mathcal{M}$-set in a mathematical sense. The procedure of an extension of a product of two complex numbers is just an assumption made by White. Moreover, applying rotations in different order, from different axes and different angles, one obtains different results (many variations appeared based on this phenomenon). However, this geometrical object has become the new icon among the fractal community, after the original $\mathcal{M}$-set.

Further studies of Nylander gave the higher-order extension of (1.11)–(1.12), which is given by

$$\{\rho, \phi, \theta\}^{p} = \{\rho^{p}, p\phi, p\theta\}, \tag{1.13}$$

$$\{\rho, \phi, \theta\}^{p} = \rho^{p} \{\cos(p\theta)\cos(p\phi), \sin(p\theta)\cos(p\phi), -\sin(p\phi)\}, \tag{1.14}$$

where $\rho = \sqrt{x^2 + y^2 + z^2}$, $\phi = \arctan(y/x)$, and $\theta = \arctan\left(\sqrt{x^2 + y^2}/z\right)$.

This type of geometric structure achieves the name *Mandelbulbs*, due to its relation to the $\mathcal{M}$-set and similarity to a bulb (especially for higher-order structures). Probably the most popular graphical representation of these objects is a Mandelbulb of the order 8. Several Mandelbulbs generated using (1.14) are presented in Figure 1.23. Note that the Mandelbulb of order 2 presented in Figure 1.23 (note that z denotes there an iterated element — do not confuse with a Cartesian coordinate z) is quite different with respect to those presented in Figure 1.22. This is because the distance estimation in this case was calculated based on analytic formulas that are used quite often in 3-D fractal generation software in order to speed up the algorithm. However, the drawback of this algorithm is lower accuracy, especially in the case of lower-order 3-D $\mathcal{M}$-sets.

The analysis of the case when $p \to \infty$ gives similar results as in the case of $\mathcal{M}$-set on the $\mathbb{C}$-plane, i.e., when $p \to \infty$ the parameter c becomes insignificant and a shape of the resulting object tends to a sphere [66].

Several other research studies have been performed to date in order to obtain complex triplets with properties similar to $\mathbb{C}$ algebra or visualize the 3-D $\mathcal{M}$-set (see, e.g., [21, 88]), but without much success.

Obviously, the original White-Nylander formula (1.14) has been modified by numerous scientists and fractal enthusiasts. As a result, many interesting shapes with a fractal hull were obtained. One can mention the so-called *Juliabulb* family obtained by applying (1.14) to $\mathcal{J}$-sets, 3-D $\mathcal{T}$-sets, *Mandelbox* family with multifractal properties discovered by Tom Lowe, and many others. Several examples of such modifications are shown in Figures 1.24 and 1.25. As one can observe, the Mandelboxes do not contain smaller copies of themselves. They are Cantor dust-type fractals, and unlike the $\mathcal{M}$-set, they can be generalized to higher-dimensional spaces.

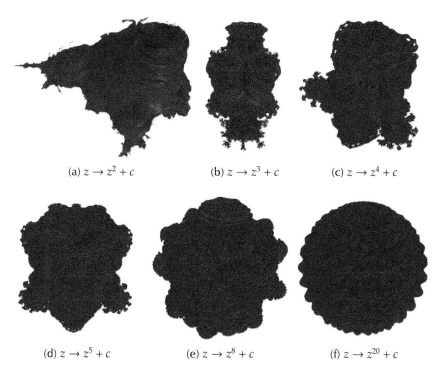

(a) $z \rightarrow z^2 + c$ (b) $z \rightarrow z^3 + c$ (c) $z \rightarrow z^4 + c$

(d) $z \rightarrow z^5 + c$ (e) $z \rightarrow z^8 + c$ (f) $z \rightarrow z^{20} + c$

Figure 1.23: Higher-order Mandelbulbs.

Coloring of 3-D fractals is slightly different than in case of fractals on a $\mathbb{C}$-plane, because now a surface of a 3-D object is to be colored, although some of the methods described in Section 1.1.2 are also applicable for 3-D fractals (e.g., assigning colors based on the number of iterations). The simplest coloring methods available in most of 3-D fractal generating and rendering software are the gradients across the axes or orbit distance, and radial mapping with linear and nonlinear functions. More advanced coloring algorithms use an extended version of escape-time algorithm, and sometimes coloring is enriched by the translucency for the voxels that

Figure 1.24: Examples of Juliabulbs.

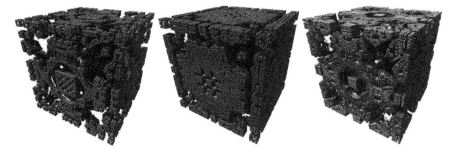

Figure 1.25: Examples of Mandelboxes.

exceed bailout value, which allows the introduction of pleasantly looking ambient occlusion effects. Additionally, the fog and lighting effects are applied in several rendering algorithms of 3-D fractals.

1.4 WHAT'S NEXT? FURTHER EXTENSIONS OF THE MANDELBROT SET

Years after Mandelbrot had discovered his set, its numerous generalizations appeared, and there are still new directions to extend classic M- and J-sets. Most of them are the higher-dimensional generalizations of M- and J-sets defined on the $\mathbb{C}$-plane, therefore, such fractals can be classified as *hypercomplex fractals*. In the next chapters, the possible ways of extensions of M- and J-sets are discussed in detail. But now, let us briefly define these directions of such extensions.

The first extension of M- and J-sets is related to the algebra of quaternions $\mathbb{H}$ (the natural extension of the algebra of complex numbers $\mathbb{C}$), which was proposed by Norton [94, 95] and further developed by Gomatham et al. [43]. Such generalization gives the new class of hypercomplex fractals and a rich range of possibilities for further generalizations in terms of variations of a form of the recursive equation (1.1) for $z, c \in \mathbb{H}$, as well as generalizations with respect to the order of z, similarly as for classic $\mathcal{M}$- and $\mathcal{J}$-sets (see (1.6)). Such a generalization was described by Wang and Sun [120]. The next hypercomplex algebra generalized the $\mathbb{H}$-algebra is the algebra of octonions $\mathbb{O}$, and the generalized $\mathcal{M}$- and $\mathcal{J}$-sets in terms of octonions were described in [44, 45, 46], shortly after introducing quaternionic $\mathcal{M}$- and $\mathcal{J}$-sets. Since further generalizations in terms of dimensionality of hypercomplex vector spaces were not possible, the Clifford algebras were used for construction of higher-dimensional hypercomplex $\mathcal{M}$- and $\mathcal{J}$-sets. The first studies in this matter were performed by Dixon et al. [35]. Based on this work, Wang and Jin [118] presented an extended mathematical analysis of such a generalization. With the help of hypercomplex algebras and advanced computer graphics tools, the new fractal beasts were born. These beasts are the subject of the second chapter of this book. However, there is also a group of vector spaces in which the mathematical rules are not the same as in the previously discussed spaces, i.e., the fractals in these spaces are degenerated or cannot be constructed at all. The analysis of such degenerated structures is performed, and based on results of this analysis, it is possible to conclude which spaces can be used for construction of higher-dimensional hypercomplex fractals, and which of them cannot. In other words, this chapter completes the list of algebras that are appropriate for construction of generalized $\mathcal{M}$- and $\mathcal{J}$-sets.

And finally, the last considered group of spaces where the fractals based on $\mathcal{M}$- and $\mathcal{J}$-sets can be constructed are the spaces constructed from the tensor product of algebras. The first study on generalized bicomplex $\mathbb{C} \otimes \mathbb{C}$ $\mathcal{M}$- and $\mathcal{J}$-sets was presented by Rochon [107]. The bicomplex $\mathcal{M}$- and $\mathcal{J}$-sets have unique properties. The deep studies on these properties were performed by the authors of [82, 119]. Another interesting study on polynomial version of $\mathcal{M}$-set for bicomplex numbers was presented by Zireh [123]. The higher-dimensional generalizations of $\mathcal{M}$- and $\mathcal{J}$-sets were studied for quaternion and biquaternion $\mathbb{C} \otimes \mathbb{H}$ algebras [12], as well as for octonion [71] and bioctonion $\mathbb{C} \otimes \mathbb{O}$ algebras [62]. Another type of generalized $\mathcal{M}$- and $\mathcal{J}$-sets was presented in [39], where the authors considered multicomplex numbers for construction of $\mathcal{M}$- and $\mathcal{J}$-sets. Naturally, one can easily obtain multihypercomplex vector spaces and use them for construction of appropriate $\mathcal{M}$- and $\mathcal{J}$-sets. Additionally, all of these algebras can be generalized with respect to the polynomial order. A group of tensor product algebras creates a possibility to construct a lot of new classes of hypercomplex and multihypercomplex $\mathcal{M}$- and $\mathcal{J}$-sets with fascinating properties and geometry. Due to how they are constructed, let us call them fractal mutants. They are described in the third chapter.

In the following chapters, both mathematical and visualization aspects are analyzed. For every considered vector space, a reader will find a detailed mathematical description as well as many rendered aesthetic fractal objects with a discussion on their visualization features.

CHAPTER 2

Fractal Beasts in Terms of Normed Division and Clifford Algebras

In the previous chapter, we discussed the history of appearance and fundamentals of fractals on a $\mathbb{C}$-plane, and different types of their generalizations. However, we did not discuss their extension into higher-dimensional spaces (except the fractal-like 3-D $\mathcal{M}$-set). In this chapter, we will move to the world of creatures and beasts living in higher-dimensional vector spaces, which sometimes have weird, but still self-similar and very complex in a geometrical sense, shapes. In spite of their young age, such hypercomplex fractals gained great popularity among the fractal community, both scientists and computer graphics artists.

One of the approaches to constructing hypercomplex fractals uses hypercomplex vector spaces, namely, the spaces defined by the algebra of quaternions and octonions. The higher-dimensional algebras in this approach cannot be used for construction of hypercomplex fractals, which will be discussed further. The second approach assumes the construction of hypercomplex fractals using Clifford algebras, which are not limited as the normed division ones. In the next three sections, we will analyse the hypercomplex numbers and a specificity of construction of fractals based on them. And we will try to imagine these higher-dimensional beasts!

2.1 QUATERNIONIC FRACTAL SETS

2.1.1 History and Preliminaries of Quaternion Calculus

The beginning of quaternions was introduced in Section 1.3 when discussing the unsuccessful attempts of Sir William R. Hamilton to extend the complex numbers to their 3-D analogues. He spent a couple of years on this problem and later wrote to his son, "Every morning in the early part of the above-cited month, on my coming down to breakfast, your (then) little brother William Edwin, and yourself, used to ask me: 'Well, Papa, can you multiply triplets?' Whereto I was always obliged to reply, with a sad shake of the head: 'No, I can only add and subtract them.'" [5]. Fortunately, Hamilton abandoned work on this issue, and on October 16, 1843, he discovered quaternions while he walked with his wife along the Royal Canal to a meeting of the Royal Irish Academy in Dublin [5]. When the concept of quaternions took a shape in his mind, he carved the identity equation

$$i^2 = j^2 = k^2 = ijk = -1 \qquad (2.1)$$

in the stone of the Broome Bridge. This action was commemorated by a stone plaque on the northwest corner of the underside of the bridge that exhibits the famous equation, which has been there until now. For more details, readers are referred to [28]. Readers interested in more information on wide applicability and several curiosities with quaternions is referred to [50].

The skew field of quaternions $\mathbb{H}$ ($\mathbb{H}$ after his discoverer — Hamilton) is a 4-algebra with basis $e_0 = 1$, $e_1 = i$, $e_2 = j$, and $e_3 = k$, where i, j, and k are the imaginary units. The set of quaternions has the form

$$\mathbb{H} := \left\{ a_1 + a_2 i + a_3 j + a_4 k \,\middle|\, a_n \in \mathbb{R} \right\}. \qquad (2.2)$$

Quaternions can be regarded as vectors in $\mathbb{R}^4$. The main difference in properties of quaternions with respect to complex numbers is that quaternions do not fulfill the condition (1.2) for $x, y \in \mathbb{H}$, i.e., they are not commutative. This means that the rules of multiplication of two quaternions are somewhat different than the rules to which we are accustomed to, e.g. $xy = z$, while $yx = -z$, for $x, y, z \in \mathbb{H}$. Since in the problem of construction of hypercomplex fractals we use addition and multiplication operations only (see (1.1)), let us analyze these operations in terms of quaternions. Considering $\mathbb{H}$ as a set of quadruples: $\mathbb{H} = \{(a_1, a_2, a_3, a_4) | a_1, \dots, a_4 \in \mathbb{R}\}$, the basis elements can be presented

TABLE 2.1
Multiplication Table for
Quaternions.

×	1	i	j	k
1	1	i	j	k
i	i	-1	k	$-j$
j	j	$-k$	-1	i
k	k	j	$-i$	-1

as $1 = (1,0,0,0)$, $i = (0,1,0,0)$, $j = (0,0,1,0)$, and $k = (0,0,0,1)$. The addition is performed element-wise and has a form

$$(a_1,a_2,a_3,a_4) + (b_1,b_2,b_3,b_4) = (a_1 + b_1, a_2 + b_2, a_3 + b_3, a_4 + b_4). \quad (2.3)$$

The multiplication of two quaternions is ruled by a distributive law that can be realized using the special multiplication table (see Table 2.1), and has a form

$$\begin{aligned}
(a_1,a_2,a_3,a_4)(b_1,b_2,b_3,b_4) = (&a_1b_1 - a_2b_2 - a_3b_3 - a_4b_4, \\
&a_1b_2 + a_2b_1 + a_3b_4 - a_4b_3, \\
&a_1b_3 - a_2b_4 + a_3b_1 + a_4b_2, \\
&a_1b_4 + a_2b_3 - a_3b_2 - a_4b_1). \quad (2.4)
\end{aligned}$$

The relation of quaternions to complex numbers can also be considered through the Cayley–Dickson construction, which means that quaternions can be represented by pairs of complex numbers: $(a_1 + a_2i) + (a_3 + a_4i)\,j$, i.e., $\mathbb{H} = \mathbb{C} \oplus \mathbb{C}$.

2.1.2 Quaternionic Julia and Mandelbrot Sets

Considering the relations between quaternions and complex numbers described previously, the formulations applicable to fractal sets on a $\mathbb{C}$-plane can be extended to their analogues in an $\mathbb{H}$-space. Two sets of prisoners and escapees also exist in an $\mathbb{H}$-space, and the boundary between these sets can be interpreted as a boundary of basin of attraction in the hypercomplex 4-space or the quaternionic $\mathcal{J}$-set – $\mathcal{J}^{\mathbb{H}}$. Similarly, as for $\mathcal{J}$-sets on a complex plane – $\mathcal{J}^{\mathbb{C}}$-sets, the bailout condition is preserved for $\mathcal{J}^{\mathbb{H}}$-sets with the bailout value equal 2, since $|z_0| > 2$ and $2 \notin \mathcal{J}^{\mathbb{H}}$, $z_0 \in \mathbb{H}$. A reader interested in hypercomplex dynamics of such systems (and several subtle differences with respect to $\mathcal{J}^{\mathbb{C}}$-sets) is referred to [72, 99, 100].

In fact, since the resulting sets are four-dimensional they cannot be graphically presented in a direct way due to the fact that humans' perception is limited to only three dimensions. Therefore, the visualizations of 4-D objects can be presented as 3-D "shadows" of these objects. In order to understand this limitation, one can recall a planar projection of a 3-D object. Suppose we have a 3-D object illuminated from the one side. The shadow of this object on a wall is its silhouette or, technically speaking, its planar projection. Now, let us extend this operation to the case of 4-D→3-D projection, i.e. in this case, the "shadow on a wall" is a 3-D spatial projection of a 4-D illuminated object. We can also follow the suggestion of Hart et al. [53] and exclude the fourth element of $\mathbb{H}$ from consideration, which gives us a 3-D subspace of a $\mathbb{H}$-space and a 3-D picture of a $\mathcal{J}^{\mathbb{H}}$-set. The exemplary $\mathcal{J}^{\mathbb{H}}$-sets are presented in Figure 2.1. As mentioned before, $\mathcal{J}^{\mathbb{C}}$- and $\mathcal{J}^{\mathbb{H}}$-sets are related, and this relation is clearly visible when coefficients at a_3 and a_4 in (2.2) of the quaternionic constant c in the recurrence quadratic polynomial (1.1) equal zero. If the 3-D projection of a $\mathcal{J}^{\mathbb{H}}$-set (see Figure 2.2a) is cross-sectioned along the $\mathbb{C}$-plane, one can achieve the related $\mathcal{J}^{\mathbb{C}}$-set in the plane of cross-section (see Figure 2.2b) — compare the resulting cross-section with Figure 1.7a. For clarity of representation, we introduce here the symbol $\mathcal{J}^{\mathbb{H}}(a_1, a_2, a_3, a_4)$, which denotes $\mathcal{J}^{\mathbb{H}}$-set with coefficients of c, $c \in \mathbb{H}$, given in brackets. In other words, if $c \in \mathbb{C}$ instead of $c \in \mathbb{H}$, the resulting $\mathcal{J}^{\mathbb{H}}$-set is trivial, which was proven by Bogush et al. [12]. Several other cross-sectioned analogues of complex fractal sets in the $\mathbb{H}$-space are presented in Figure 2.3.

Nowadays, for graphical representation of quaternionic fractal sets, the ray-tracing approach is used in most computer programs to generate and render fractals due to the ability to generate very realistic renderings, including complex lighting, transparency, reflection, and refraction effects. The other previously used techniques, like boundary tracking or inverse iteration algorithm, are rarely used to render quaternionic fractal sets. The history of developing of visualization techniques of quaternionic fractals can be found in [51, 52, 109]. The theory and modern techniques of visualizing quaternions can also be found in [50].

Using the analogy with 3-D "shadows," we can also define the slices of 4-D objects. The slice of a 3-D object is planar. One can imagine the process of performing slicing of a 3-D object as follows: first, cut 3-D object into two pieces, then put one of these pieces into a paint, and, finally, make an imprint on a paper. In the case of a 4-D object, we have a 3-D cutting hyperplane, which allows obtaining 3-D slices of the object.

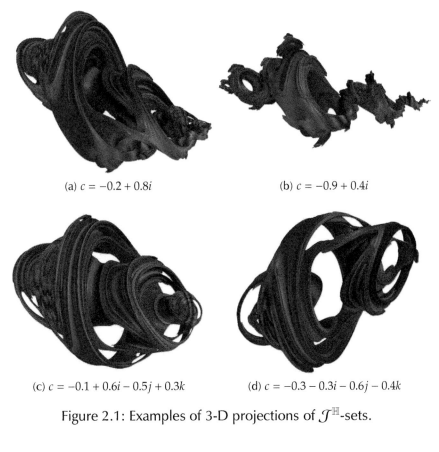

(a) $c = -0.2 + 0.8i$

(b) $c = -0.9 + 0.4i$

(c) $c = -0.1 + 0.6i - 0.5j + 0.3k$

(d) $c = -0.3 - 0.3i - 0.6j - 0.4k$

Figure 2.1: Examples of 3-D projections of $\mathcal{J}^{\mathbb{H}}$-sets.

(a)

(b)

Figure 2.2: 3-D projection of $\mathcal{J}^{\mathbb{H}}(-0.5, 0.5, 0, 0)$ and its cross-section along the $\mathbb{C}$-plane.

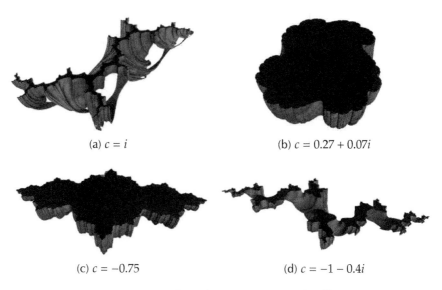

(a) $c = i$

(b) $c = 0.27 + 0.07i$

(c) $c = -0.75$

(d) $c = -1 - 0.4i$

Figure 2.3: Examples of cross-sectioned $\mathcal{J}^{\mathbb{H}}$-sets.

Obviously, when we make such a slice, one of coordinates of a 4-space (i.e., the element of a quaternion) is constant. Of course, in this case, our image of the 4-D object will be limited to a single slice. In order to get greater insight on how this object looks in 4-space, one needs to generate a sequence of 3-D slices with a discrete stepping of a position of a cutting hyperplane. The examples of such a procedure are presented in Figure 2.4 for slicing along axis of reals and i-axis, and in Figure 2.5 for slicing along i-axis and j-axis.

An interesting observation can be made when rotating $\mathcal{J}^{\mathbb{H}}$-set. Since a rotation is performed in a 4-space a 3-D projection of a $\mathcal{J}^{\mathbb{H}}$-set acquires completely different shapes and topology [53]. This behavior is resulted by the "shadow effect," i.e., when we rotate a 3-D object with complex geometry, its shadow on a wall is different. The same happens when we rotate a 4-D object and observe its 3-D shadow [63]. However, the rotation of $\mathcal{J}^{\mathbb{H}}$-sets in an $\mathbb{H}$-space is not so obvious due to the lack of commutativity of a multiplied pair of quaternions. The rotations in 4-D can be considered as a multiplication of a point in a 4-space $\mathbf{q}$ by a unit quaternion (which can be obtained by dividing the nonzero quaternion by its norm: $\mathbf{q}/\|\mathbf{q}\|$). Depending on the direction of rotation, the left-

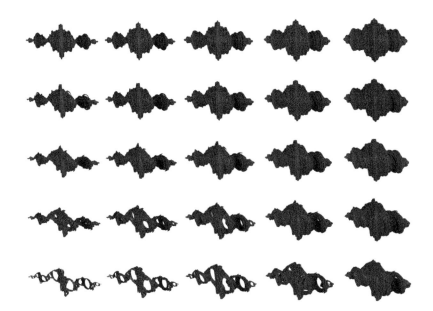

Figure 2.4: Slices of $\mathcal{J}^{\mathbb{H}}(-1, 0.1, 0, 0)$ along axis of reals (from -1 to -0.6) and i-axis (from 0.1 to 0.5) with a step of 0.1.

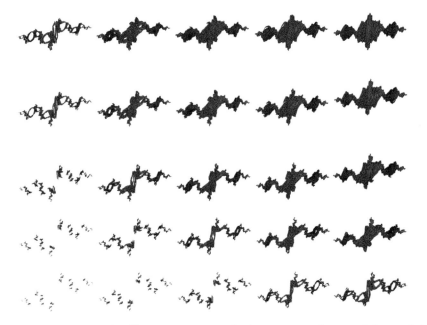

Figure 2.5: Slices of $\mathcal{J}^{\mathbb{H}}(-1, -0.5, 0.1, 0)$ along i-axis (from -0.5 to -0.1) and j-axis (from 0.1 to 0.5) with a step of 0.1.

multiplication by a unit quaternion $q_1 = b_1 + b_2 i + b_3 j + b_4 k$

$$
\begin{pmatrix} a'_1 \\ a'_2 \\ a'_3 \\ a'_4 \end{pmatrix} = \begin{pmatrix} b_1 & -b_2 & -b_3 & -b_4 \\ b_2 & b_1 & -b_4 & b_3 \\ b_3 & b_4 & b_1 & -b_2 \\ b_4 & -b_3 & b_2 & b_1 \end{pmatrix} \begin{pmatrix} a_1 \\ a_2 \\ a_3 \\ a_4 \end{pmatrix},
\tag{2.5}
$$

or the right-multiplication by a unit quaternion $q_2 = c_1 + c_2 i + c_3 j + c_4 k$

$$
\begin{pmatrix} a'_1 \\ a'_2 \\ a'_3 \\ a'_4 \end{pmatrix} = \begin{pmatrix} c_1 & -c_2 & -c_3 & -c_4 \\ c_2 & c_1 & c_4 & -c_3 \\ c_3 & -c_4 & c_1 & c_2 \\ c_4 & c_3 & -c_2 & c_1 \end{pmatrix} \begin{pmatrix} a_1 \\ a_2 \\ a_3 \\ a_4 \end{pmatrix},
\tag{2.6}
$$

should be performed. This can be simplified to the symbolic form

$$
q' = q_1 q q_2.
\tag{2.7}
$$

This rotation, however, is still performed in 4-space, and the explicit result of rotation in 3-space is difficult to obtain. Therefore, it is essential to apply the 3-D rotation matrix to the point in 4-space mapped into 3-space (given by $\{x, y, z\}$) using the Hamilton–Cayley formula [28]

$$
\begin{pmatrix} x' \\ y' \\ z' \end{pmatrix} = \begin{pmatrix} d_1^2 + d_2^2 - d_3^2 - d_4^2 & 2(d_2 d_3 - d_1 d_4) & 2(d_2 d_4 + d_1 d_3) \\ 2(d_2 d_3 + d_1 d_4) & d_1^2 - d_2^2 + d_3^2 - d_4^2 & 2(d_3 d_4 - d_1 d_2) \\ 2(d_2 d_4 - d_1 d_3) & 2(d_3 d_4 + d_1 d_2) & d_1^2 - d_2^2 - d_3^2 + d_4^2 \end{pmatrix} \begin{pmatrix} x \\ y \\ z \end{pmatrix},
\tag{2.8}
$$

where d_n's are the multipliers of the elements of a quaternion.

An even simpler method used in most computer programs for generation of $\mathcal{J}^{\mathbb{H}}$-sets considers (2.7) as a rotation performed in a 3-space by an angle θ around the axis ξ, so the rotation of a point r around the axis ξ can be expressed as $p' = gpg^{-1}$, where p is a quaternion containing r,

$$
g = \cos(\theta/2) + \sin(\theta/2)\,\xi.
\tag{2.9}
$$

Here, ξ defines a vector direction in a 3-space. This has a consequence that any quaternion can be rotated into a complex number by conjugation with the appropriate unit quaternion [30].

Sometimes, it is convenient to define $\mathcal{J}^{\mathbb{H}}$-sets in hyperspherical coordinates. The Euclidean coordinates of an $\mathbb{H}$-space (2.2) are related to hyperspherical coordinates as follows:

$$
a_1 = A \cos\theta_1, \quad a_2 = A \sin\theta_1 \cos\theta_2,
$$
$$
a_3 = A \sin\theta_1 \sin\theta_2 \cos\theta_3, \quad a_4 = A \sin\theta_1 \sin\theta_2 \sin\theta_3,
\tag{2.10}
$$

where $A = \sqrt{\sum_n a_n^2}$, θ_1 is the amplitude of quaternion, and θ_2 and θ_3 are the latitude and longitude, respectively. In analogy to the Euclidean coordinates, θ_1, θ_2 and θ_3 are the angular coordinates. Similarly, as for fractal sets on a $\mathbb{C}$-plane, the Mandelbrot set defined on a $\mathbb{C}$-plane, $\mathcal{M}^\mathbb{C}$-set, can be extended to quaternions. However, it is difficult to fully describe the $\mathcal{M}^\mathbb{H}$-set when the connected $\mathcal{J}^\mathbb{H}$-sets for $f(z) = z^2 + c$, $z, c \in \mathbb{H}$ for $c \notin \mathcal{M}^\mathbb{H}$. Considering the definition of the $\mathcal{M}^\mathbb{H}$-set with respect to $f(z)$ [122],

$$\mathcal{M}^\mathbb{H} = \left\{ c \in \mathbb{H} \mid f^{(s)}(z) \not\to \infty \text{ as } s \to \infty \right\}. \tag{2.11}$$

In these cases, the connected $\mathcal{J}^\mathbb{H}$-sets are infinite disjointed unions of sets, since the derivative of $f(z)$ is not defined and there are no critical points. Therefore, there is no analogue to the classification theorem for $\mathcal{J}^\mathbb{C}$-sets in terms of the $\mathcal{M}^\mathbb{C}$-set in quaternions [30]. In order to evaluate $\mathcal{J}^\mathbb{H}$, we assume the seed of $z_0 = 0$, hoping that by iteration really all cycles are reached [122]. The $\mathcal{M}^\mathbb{H}$-set is usually presented using the above-described technique of excluding a k-element from consideration. Obviously, the resulting object is trivial (see Figure 2.6), and, in fact, represents a rotation of the $\mathcal{M}^\mathbb{C}$-set around the axis of reals. This fact was proven by Zeitler [122] and known as *the extended apple theorem*. Yet, as Barrallo [6] emphasizes, the appropriate choice of cutting hyperplane orientation can give more interesting 3-D slices of the $\mathcal{M}^\mathbb{H}$-set.

Figure 2.6: 3-D projection of the $\mathcal{M}^\mathbb{H}$-set along the $\mathbb{C}$-plane.

Obviously, $\mathcal{J}$-sets in $\mathbb{H}$ can be considered in terms of higher-order polynomials that describe them. Let us first consider the recursive equation in the form (1.5), when $z, c \in \mathbb{H}$, $p \in \mathbb{Z}$. The resulting slices for various values of p are presented in Figure 2.7, similarly to the case of $\mathcal{J}$-sets on a $\mathbb{C}$-plane presented in Figure 1.7. One can observe that the 3-D projections of $\mathcal{J}_p^{\mathbb{H}}(-0.5, 0.5, 0, 0)$ for increasing p tend to a 2-sphere. Indeed, when $p \to \infty$, the shape of $\mathcal{J}_p^{\mathbb{H}}(a_1, a_2, a_3, a_4)$ tends to a 3-sphere for arbitrary a_n, which was proven in [66].

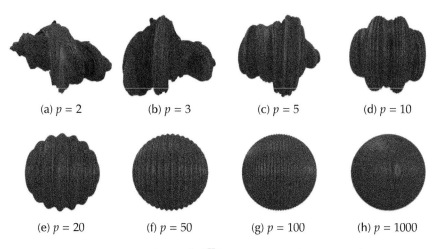

(a) $p = 2$ (b) $p = 3$ (c) $p = 5$ (d) $p = 10$

(e) $p = 20$ (f) $p = 50$ (g) $p = 100$ (h) $p = 1000$

Figure 2.7: Tending of $\mathcal{J}_p^{\mathbb{H}}(-0.5, 0.5, 0, 0)$ to a 2-sphere.

The mystery of quaternionic fractal sets does not end at a class of polynomials of the form (1.5). The 4-space gives even more possibilities of creating amazing shapes by modified recursive equations only. But only the few examples of such modifications can be found in the literature, e.g., in [109], the author visualized transcendental quaternionic functions and quasi-$\mathcal{M}^{\mathbb{H}}$-sets with modified critical points; Halayka [49] presented several visually interesting $\mathcal{J}^{\mathbb{H}}$-sets based mostly on transcendental functions; and Sun et al. [116] presented $\mathcal{J}^{\mathbb{H}}$- and $\mathcal{M}^{\mathbb{H}}$-sets perturbed by dynamical noises. Several examples of visually interesting variations of $\mathcal{J}^{\mathbb{H}}$-sets are presented in Figures 2.8 and 2.9.

(a) $z \to z^2 - c/z$,
$c = 0.5 + 0.3i - 0.1j$

(b) $z \to cz^2 - cz^{-2}$,
$c = 0.3 + 0.3i - 0.5j - 0.2k$

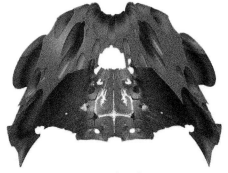

(c) $z \to cz^{-2} - c^2 z + c$,
$c = 0.3i - 0.3j$

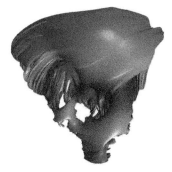

(d) $z \to \ln(c)/z - z^{-4}$,
$c = 0.3 - 1.9i + 1.1j - 0.2k$

(e) $z \to c^2/\ln(c) + z^{-4}$,
$c = 0.55 + 0.4i - 0.5j + 0.5k$

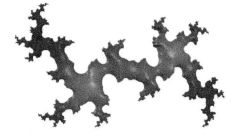

(f) $z \to z/\ln(c) - (zc)^2$,
$c = 0.6 + 0.6i - 0.4j$

Figure 2.8: Variations of $\mathcal{J}^{\mathbb{H}}$-sets (part 1 of 2).

(a) $z \to 1/\ln(c) + \ln(z^2 + c)$,
$c = 0.4 + 0.4i - 0.6j$

(b) $z \to (1 - z - c)^2 - (zc)^2$,
$c = 0.5 - 0.3i + 0.3j - 0.1k$

(c) $z \to 1 - c + (\ln(z) + c)^2$,
$c = 0.6 - 0.29j$

(d) $z \to (z/c)^4 - z/c$,
$c = 0.27 + 0.35i - 0.5j$

(e) $z \to c\ln(c) + z^2/c$,
$c = 0.75 + 0.67i - 0.34j$

(f) $z \to 1.6/\ln(c) + \ln(z^2 + c)$,
$c = 0.1 + 1i + 0.1j + 0.1k$

Figure 2.9: Variations of $\mathcal{J}^{\mathbb{H}}$-sets (part 2 of 2).

2.2 OCTONIONIC FRACTAL SETS

2.2.1 History and Preliminaries of Octonions

The history of octonions is not less interesting than the history of quaternions. In fact, this history is a continuation of the story of Hamilton's discovery of quaternions. Despite the fact that the octonions are known also as *Cayley numbers* after British mathematician Arthur Cayley, they were first discovered by John T. Graves, Hamilton's friend from college [5]. After discovering quaternions on October 16, 1843, Hamilton wrote a letter to Graves with their description. Graves replied to this letter suggesting to expand quaternions to higher-dimensional hypercomplex vector spaces. Since Hamilton was obsessed with his quaternions, Graves started working on such an extension on his own. And on December 26, 1843, he wrote a letter to Hamilton, in which he described a new 8-algebra that he called the *octaves* (however, this name was supplanted by the "Cayley numbers"), which was a direct extension of quaternions. Hamilton suggested that Graves publish his discovery, but being busy with his work on quaternions, he put it on the back burner. In July 1844, Hamilton wrote to Graves about his finding on octonions — in contrast to quaternions, they are not associative.

In the meantime, Arthur Cayley, freshly graduated from Cambridge, had been working with quaternions since Hamilton's announcement. In March 1845, he published a paper [16], where, among other things, he presented a brief description on octonions. Both Graves and Hamilton published notes in well recognizable mathematical journal claiming Graves' priority in discovering the octonions, but it was too late — the octonions became known as "Cayley numbers" [5]. For more information on the details of this story, a reader is referred to [5, 50].

The octonions $\mathbb{O}$ is an 8-algebra with the basis of $e_0 = 1$, $e_1 = i$, $e_2 = j$, $e_3 = k$, $e_4 = l$, $e_5 = r$, $e_6 = s$, $e_7 = t$, $e_n^2 = -1$, $n = 1,\ldots,7$, which lose both commutativity (1.2) and associativity (1.3) conditions, i.e., this algebra is anticommutative and antiassociative. The only condition that $\mathbb{O}$-algebra holds is the alternativity (1.4). The set of octonions has the following symbolic form:

$$\mathbb{O} := \left\{ a_1 + a_2 i + a_3 j + a_4 k + a_5 l + a_6 r + a_7 s + a_8 t \,\middle|\, a_n \in \mathbb{R} \right\}. \qquad (2.12)$$

Similarly, as for the $\mathbb{H}$-algebra, the addition of two octonions is element-wise, i.e., considering $\mathbb{O}$ as a set of octuples in $\mathbb{R}^8$, the basis elements can be presented as

$$e_0 = (1,0,0,0,0,0,0,0), \ e_1 = (0,1,0,0,0,0,0,0),$$
$$e_2 = (0,0,1,0,0,0,0,0), \ e_3 = (0,0,0,1,0,0,0,0),$$
$$e_4 = (0,0,0,0,1,0,0,0), \ e_5 = (0,0,0,0,0,1,0,0),$$
$$e_6 = (0,0,0,0,0,0,1,0), \ e_7 = (0,0,0,0,0,0,0,1); \tag{2.13}$$

and then, the addition has a form

$$(a_1,\ldots,a_8) + (b_1,\ldots,b_8) = (a_1 + b_1,\ldots,a_8 + b_8). \tag{2.14}$$

Again, similarly to $\mathbb{H}$-algebra, the multiplication is realized using the multiplication table (see Table 2.2 after Cayley [16]). The table can be expressed by the following relations [5, 112]:

$$e_m e_n = -\delta_{mn} e_0 + \epsilon_{mnp} e_p, \tag{2.15}$$

where δ is the Kronecker delta, $\epsilon_{mnp} = 1$, when mnp takes the values of $123, 145, 176, 246, 257, 347, 365$,

$$e_m e_0 = e_0 e_m = e_m, \ e_0 e_0 = e_0, \tag{2.16}$$

$$e_m e_n = -e_n e_m \text{ for } m \neq n, \tag{2.17}$$

and two additional identities called *index cycling*,

$$e_m e_n = e_p \Rightarrow e_{m+1} e_{n+1} = e_{p+1}, \tag{2.18}$$

and *index doubling*,

$$e_m e_n = e_p \Rightarrow e_{2m} e_{2n} = e_{2p}. \tag{2.19}$$

Even considering relations and identities (2.15)–(2.19), the multiplication matrix for octonions is still difficult to remember. Fortunately, Freudenthal [38] proposed a mnemonic based on the Fano plane concept (Figure 2.10). This structure contains 7 nodes and lines connecting them. Every path contains 3 nodes that define the multiplication (see Table 2.2 and Figure 2.10). When following the direction of the arrows, the product is positive, e.g., $e_3 e_4 = e_7$. While when following the opposite direction, the product is negative, e.g., $e_4 e_3 = -e_7$. Each of the 7 lines (i.e., any two elements) generates a subalgebra of $\mathbb{O}$ isomorphic to all other normed division algebras, namely $\mathbb{R}$, $\mathbb{C}$, and $\mathbb{H}$.

The octonions are related to $\mathbb{H}$- and $\mathbb{C}$-algebras through the Cayley–Dickson construction, which means that octonions can be represented by pairs of quaternions: $(a_1 + a_2 i + a_3 j + a_4 k) + (a_5 + a_6 i + a_7 j + a_8 k) l$.

TABLE 2.2   Multiplication Table for Octonions.

×	e_0	e_1	e_2	e_3	e_4	e_5	e_6	e_7
e_0	e_0	e_1	e_2	e_3	e_4	e_5	e_6	e_7
e_1	e_1	$-e_0$	e_3	$-e_2$	e_5	$-e_4$	$-e_7$	e_6
e_2	e_2	$-e_3$	$-e_0$	e_1	e_6	e_7	$-e_4$	$-e_5$
e_3	e_3	e_2	$-e_1$	$-e_0$	e_7	$-e_6$	e_5	$-e_4$
e_4	e_4	$-e_5$	$-e_6$	$-e_7$	$-e_0$	e_1	e_2	e_3
e_5	e_5	e_4	$-e_7$	e_6	$-e_1$	$-e_0$	$-e_3$	e_2
e_6	e_6	e_7	e_4	$-e_5$	$-e_2$	e_3	$-e_0$	$-e_1$
e_7	e_7	$-e_6$	e_5	e_4	$-e_3$	$-e_2$	e_1	$-e_0$

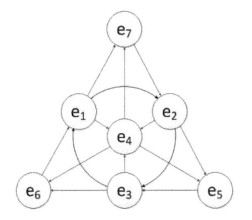

Figure 2.10: Fano plane — a mnemonic for multiplication of octonions.

2.2.2   Octonionic Julia and Mandelbrot Sets

In the case of octonions, the formulation of the recursive formula for $\mathcal{J}$- and $\mathcal{M}$-sets defined in octonions, $\mathcal{J}^{\mathbb{O}}$- and $\mathcal{M}^{\mathbb{O}}$-sets, is much more complicated than for their lower-dimensional analogues defined in quaternions and complex numbers. This is because of the nonassociativity of the $\mathbb{O}$-algebra. Following this, the recursive equation (1.1) takes a form [1]:

$$z \rightarrow (zo)(\bar{o}z) + c, \quad z, o, c \in \mathbb{O}, \tag{2.20}$$

where o is a constant unit octonion that has a form of (2.12) for $a_n = 1$, and $\bar{o}$ is a hypercomplex conjugate of o, i.e., $\bar{o} \equiv a_1 - a_2 i - a_3 j - a_4 k - a_5 l - a_6 r - a_7 s - a_8 t$. The multiplication in (2.20) is performed using a concept of octonionic cross-product proposed by Cederwall and Preitschopf [17]. Note that (2.20) is reduced to the form given by (1.1) for $z, c \in \mathbb{H}$.

The nonassociativity drives the properties of $\mathcal{J}^{\mathbb{O}}$- and $\mathcal{M}^{\mathbb{O}}$-sets, which was the subject of intensive investigations by Griffin and Joshi [44, 45, 46], and later by Kricker and Joshi [69], where they studied dynamics of these sets, and found that they reveal chaotic and even hyperchaotic regimes. The attractive structures of $\mathcal{J}^{\mathbb{O}}$- and $\mathcal{M}^{\mathbb{O}}$-sets, as well as the fixed points, depend on the bifurcation parameter, which is related to nonassociativity of the $\mathbb{O}$-algebra, as observed in [69]. Though, Chaitin-Chatelin [18] demonstrated that (2.20) may generate trivial solutions, similarly th $\mathcal{J}^{\mathbb{H}}$- and $\mathcal{M}^{\mathbb{H}}$-sets. This means that when $a_n = 0$ for $n = 3, \ldots, 8$ in (2.12), i.e., $c \in \mathbb{C}$ instead of $c \in \mathbb{O}$, the resulting set can be obtained by rotation of its analogue in a $\mathbb{C}$-plane.

The visualization of octonionic $\mathcal{J}^{\mathbb{O}}$- and $\mathcal{M}^{\mathbb{O}}$-sets is a serious problem, since having only thee geometrical dimensions (naturally resulting from humans' perception), the visualization of an 8-D object results in a huge loss of information. First of all, there is no way to visualize the 8-D object entirely. Thus, several tricky techniques were developed for this purpose. The most obvious way is to visualize 2-D or 3-D slices of an 8-D object in various planes, similar to quaternionic fractal sets (see Figures 2.4 and 2.5). An interesting approach to overcome this problem was presented by Carter [15]. The hypercomplex spaces are divided following the Cayley–Dickson construction onto quaternionic 4-spaces, and then the resulting quaternions are mapped on a 3-space separately and composed in a single image using the depths property of this 3-space. In the case of visualization of a quaternionic fractal set by 2-D cross-sections, 28 combinations of elements of $\mathbb{O}$ that form cutting planes should be considered. Some of the exemplary sets of 2-D cross-sections of the quadratic $\mathcal{J}^{\mathbb{O}}$-set described by (2.20) are presented in Figures 2.11–2.14. In each case, we start from $\mathcal{J}^{\mathbb{O}}(0,0,0,0,0,0,0,0)$ and cut it by different planes.

The $\mathcal{M}^{\mathbb{O}}$-set can be described similarly as its lower-dimensional analogues by (2.11) for $z, c \in \mathbb{O}$ and, as in the case of the $\mathcal{M}^{\mathbb{H}}$-set, the derivative of $f(z)$ is undefined, since no critical points can be determined for this set. The exemplary visualization of one of cross-sections of the $\mathcal{M}_2^{\mathbb{O}}$-set is shown in Figure 2.15.

Generalizations of the quadratic $\mathcal{M}^{\mathbb{O}}$-set with respect to its polynomial degree, as well as general form of a recursive equation, is also possible. However, due to the lack of appropriate visualization of such structures, their presentation is omitted. In the same way as previously, the $\mathcal{M}_p^{\mathbb{O}}$-set tends to the 7-sphere when $p \to \infty$. The analysis and results on the generalized $\mathcal{M}^{\mathbb{O}}$-set can be found in [45].

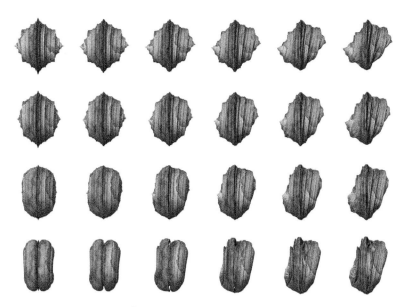

Figure 2.11: Slices of $\mathcal{J}^{\circledcirc}(0,0,0,0,0,0,0,0)$ along axis of reals (from 0 to 0.3) and i-axis (from 0 to 0.5), with a step of 0.1.

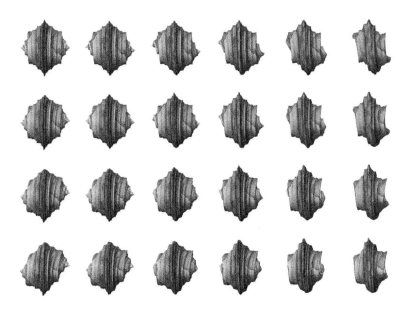

Figure 2.12: Slices of $\mathcal{J}^{\circledcirc}(0,0,0,0,0,0,0,0)$ along j-axis (from 0 to 0.3) and k-axis (from 0 to 0.5), with a step of 0.1.

Figure 2.13: Slices of $\mathcal{J}^{\circ}(0,0,0,0,0,0,0,0)$ along l-axis (from 0 to 0.3) and r-axis (from 0 to 0.5), with a step of 0.1.

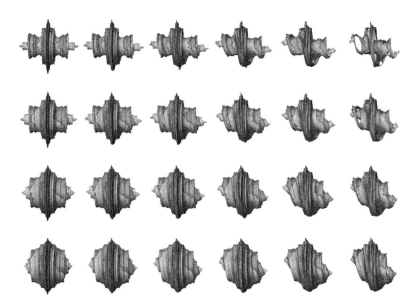

Figure 2.14: Slices of $\mathcal{J}^{\circ}(0,0,0,0,0,0,0,0)$ along axis of reals (from -0.3 to 0) and t-axis (from 0 to 0.5), with a step of 0.1.

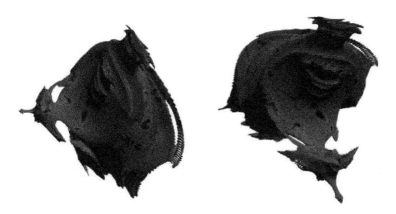

Figure 2.15: Exemplary cross-section of $\mathcal{M}_2^{\mathbb{O}}$.

2.3 IS THERE ANYTHING ALIVE IN HIGHER-DIMENSIONAL HYPERCOMPLEX SPACES?

Since we have already known that $\mathcal{M}$- and $\mathcal{J}$-sets can be generalized up to octonionic hypercomplex 8-space, the natural question is whether we can perform their further generalizations. In order to answer this question, let us get back to the story of two friends, Hamilton and Graves. Immediately after Graves had sent his letter with the description of octonions, he sent a few more letters, where he presented a concept of "2^m-ions," i.e., further generalizations of the $\mathbb{O}$-algebra. The first of them was the 16-algebra of sedenions $\mathbb{S}$ (also called the hexadecanions [19]), which should be the next normed division algebra, according to the Graves' conception. However, "he met with an unexpected hitch" and discovered that this cannot be possible [5]. In fact, the $\mathbb{O}$-algebra can be generalized considering the Cayley–Dickson construction in such a way that every next algebra doubles dimension of its vector space. In such a way, one can obtain the mentioned sedenions $\mathbb{S}$, 32-algebra of pathions $\mathbb{P}$, and so on. However all of these algebras starting from $\mathbb{S}$ are neither commutative, nor associative, nor alternative (sedenions have only a weak form of associativity known as power associativity). This means that all of the mentioned algebras are not the division algebras, thus contain zero divisors. From the point of view of constructing generalized hypercomplex $\mathcal{M}$- and $\mathcal{J}$-sets, this means that the multiplication operation cannot be performed in such space, since it is possible that the multiplication operation of two nonzero sedenions or higher-dimensional vectors gives zero as a result. This has far-reaching consequences when iterating the higher-dimensional analogue of (1.1). Let us consider a case when for c one obtains a sequence that tends to infinity (assuming the existence of a critical point at infinity). In this case, the appearance of zero in such a sequence results in changing the type of trajectory from repelling to the attractive one, and the starting point located outside the boundary of $\mathcal{M}$- and $\mathcal{J}$-sets, then belongs to the subset of prisoners, which causes a singularity of dynamical behavior of such a system. Similarly, when c belongs to the subset of prisoners, the singularity may be produced inside this subset. This implies that generalizations of $\mathcal{M}$- and $\mathcal{J}$-sets to the vector spaces constructed using the Cayley–Dickson construction that are not described by normed division algebras are not possible. The sets generated using the analogous form of (1.1) in higher-dimensional vector spaces constructed using the Cayley–Dickson construction are not $\mathcal{M}$- and $\mathcal{J}$-sets and will be called here *degenerated quasi-fractal $\mathcal{M}$- and $\mathcal{J}$-sets*. Fortunately, there is another class of algebras that generalizes previously discussed algebras, and which is the subject of this subsection – the Clifford algebras.

2.3.1 Preliminaries to Clifford Algebras

The Clifford algebras (also called the *alternions* [110] or the hypercomplex numbers [35, 118]) are named after their discoverer, William Kingdon Clifford, who worked on higher-dimensional generalizations of quaternions proposed by Hamilton. He described his invention in [23]. In general, the Clifford algebra $\mathbb{K}_n$ is a 2^n-algebra with the basis of $1, e_1, e_2, \ldots, e_n$, and its elements satisfy the conditions of associativity (1.3) and alternativity (1.4). The $\mathbb{K}_n$ is generated by n anticommuting ($xy = -yx$) imaginary units e_n following the doubling construction. One can see that in this case, several $\mathbb{K}_n$-algebras coincide with several normed division algebras, i.e., $\mathbb{K}_0 = \mathbb{R}$, $\mathbb{K}_1 = \mathbb{C}$, $\mathbb{K}_2 = \mathbb{H}$, but $\mathbb{K}_3 \neq \mathbb{O}$, since $\mathbb{O}$ is not an associative algebra. $\mathbb{K}_3$ coincides with the direct sum $\mathbb{H} \oplus \mathbb{H}$ (the direct sum of an n-algebra $\mathbb{A}$ and an m-algebra $\mathbb{B}$, with bases $e_1, e_2, \ldots, e_n$ and $f_1, f_2, \ldots, f_n$ is defined as the $(n + m)$-algebra $\mathbb{A} \oplus \mathbb{B}$, with the basis $e_1, e_2, \ldots, e_n, f_1, f_2, \ldots, f_n$, where $e_\alpha f_\beta = f_\beta e_\alpha = 0$ [110]). Further, higher-dimensional relations are as follows: $\mathbb{K}_4 = \mathbb{H} \otimes \mathbb{R}_2 = \mathbb{H}_2$, $\mathbb{K}_5 = \mathbb{R}_4 \otimes \mathbb{C} = \mathbb{C}_4$, $\mathbb{K}_6 = \mathbb{R}_2 \otimes \mathbb{R}_4 = \mathbb{H}_8$, and $\mathbb{K}_7 = \mathbb{R}_8 \otimes \mathbb{R}_8$ [110]. The form $\mathbb{A} \otimes \mathbb{B}$ denotes a tensor product of an n-algebra $\mathbb{A}$ and an m-algebra $\mathbb{B}$, with bases $e_1, e_2, \ldots, e_n$ and $f_1, f_2, \ldots, f_n$, which results in the nm-algebra $\mathbb{A} \otimes \mathbb{B}$, with basis elements $e_\alpha f_\beta = f_\beta e_\alpha$ [110]. For $n \geq 8$, the following relation occurs [5]:

$$\mathbb{K}_{n+8} \cong \mathbb{K}_n \otimes \mathbb{R}_{16}. \tag{2.21}$$

This behavior of $\mathbb{K}_{n\geq 8}$ is known as the *Bott periodicity*.

The general symbolic form of a Clifford algebra can be given by

$$\mathbb{K}_n := \{a_1 + a_2 e_1 + a_3 e_2 + \ldots + a_{n+1} e_n | a_n \in \mathbb{R}\}. \tag{2.22}$$

Since $\mathbb{K}_n$ are associative algebras, they are closed under addition and multiplication, which is crucial for performing the iteration procedure of a recursive equation of the type (1.1). The addition of two generalized hypercomplex numbers is performed, as in previous cases, element-wise:

$$(a_1 + a_2 e_1 + \ldots + a_{n+1} e_n) + (b_1 + b_2 e_1 + \ldots + b_{n+1} e_n) =$$
$$(a_1 + b_1) + (a_2 + b_2) e_1 + \ldots + (a_{n+1} + b_{n+1}) e_n, \tag{2.23}$$

while the squaring operation (multiplication by itself) yields [35]

$$(a_1 + a_2 e_1 + \ldots + a_{n+1} e_n)^2 = \left(a_1^2 - a_2^2 - \ldots - a_{n+1}^2\right) +$$
$$(2a_1 a_2) e_1 + (2a_1 a_3) e_2 + \ldots + (2a_1 a_{n+1}) e_n. \tag{2.24}$$

The definition of these operations is enough to formulate the $\mathcal{J}^{\mathbb{K}_n}$- and $\mathcal{M}^{\mathbb{K}_n}$-sets, i.e., $\mathcal{J}$- and $\mathcal{M}$-sets defined in Cliffordean algebras.

2.3.2 Cliffordean Hypercomplex Fractal Sets

Consider the function of a type

$$z \to z^p + c \text{ for } z, c \in \mathbb{K}_n, \ p \geq 2, \tag{2.25}$$

since for $-1 \leq p \leq 1$, the $\mathcal{J}_p^{\mathbb{K}_n}$ does not exist, and for $1 \leq p \leq 2 \vee p < -1$, it is a degenerated $\mathcal{J}_p^{\mathbb{K}_n}$ (see Section 1.1.3 for details), so the generalized $\mathcal{J}_p^{\mathbb{K}_n}$-set can be formulated as

$$\mathcal{J}_p^{\mathbb{K}_n}(c) = \{z \in \mathbb{K}_n | \forall_{m \in \mathbb{N}} |z_m| < B\}, \tag{2.26}$$

where $B \in \mathbb{R}$ is a bailout value usually equal to 2. Thus, the $\mathcal{M}_p^{\mathbb{K}_n}$-set is given by:

$$\mathcal{M}_p^{\mathbb{K}_n} = \left\{c \in \mathbb{K}_n | f^{(s)}(z) \not\to \infty \text{ as } s \to \infty\right\}, \tag{2.27}$$

where $f^{(s)}(\cdot)$ denotes a function composition performed s times.

As presented above, the complex numbers and the quaternions

coincide with the Clifford number systems, therefore, their description is omitted. The higher-dimensional Clifford algebras, due to the impossibility of proper visual representation, are not presented here. The only studies in the area of higher-dimensional Cliffordean fractal sets performed so far are the studies of Dixon et al. [35] and Wang and Jin [118]. The authors of both studies proposed an interesting approach to obtain higher-dimensional hypercomplex sets other than those obtained from the Cayley–Dickson construction. They selected the higher-dimensional vector space and then truncated the elements of a hypercomplex number of the form (2.22) by setting the associators a_n to zero, starting from the last one, a_n, until reaching the desired dimension. This operation allows avoiding the problems of multiplication of hypercomplex numbers, but, in fact, this process is a trivialization of a given hypercomplex vector space. Following this algorithm, we need to consider at least 8-D hypercomplex vector space in order to obtain 5-D $\mathcal{J}^{\mathbb{K}_n}$- and $\mathcal{M}^{\mathbb{K}_n}$-sets. Continuing the truncation process, we can obtain 4-D quaternionic vector space, and then 2-D complex plane. Thus, the $\mathcal{J}^{\mathbb{K}_n}$- and $\mathcal{M}^{\mathbb{K}_n}$-sets obtained from the truncation procedure are not the new classes of hypercomplex fractal sets, but the trivialized $\mathcal{J}^{\mathbb{K}_n}$- and $\mathcal{M}^{\mathbb{K}_n}$-sets of a vector space from which the truncation process started, similar to the cases described for quaternions (see Section 2.1.2) and octonions (see Section 2.2.2). Obviously, the graphical analysis presented in [35, 118] shows that the hypercomplex sets generated from (2.26) and (2.27) are graphically fractals.

2.4 THE PLACES (AND HYPERSPACES) WHERE FRACTALS CANNOT LIVE

Until now, we have analyzed the algebras and number systems that define appropriate spaces and hyperspaces for construction of $\mathcal{M}$- and $\mathcal{J}$-sets in them. The general condition for these number spaces is that they should be closed under addition and multiplication operations, thus, should fulfill at least one of three conditions: commutativity (1.2), associativity (1.3), or alternativity (1.4). Since we have already defined most of such number spaces, let us define such ones where the mentioned algebraic operations do not work well. The algebras defined in the previous chapters have a lot of derivative algebras with the same dimensions, but with varying properties.

Let us start from the derivatives of the $\mathbb{C}$-algebra. This algebra has two derivative algebras: the algebra of split complex numbers $\mathbb{C}'$, and

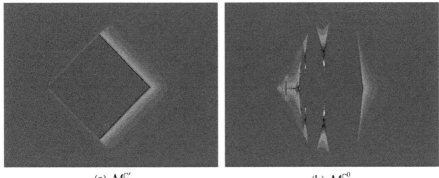

(a) $\mathcal{M}_2^{\mathbb{C}'}$ (b) $\mathcal{M}_2^{\mathbb{C}^0}$

Figure 2.16: Exemplary sets in a $\mathbb{C}'$-plane and in a $\mathbb{C}^0$-plane.

the algebra of dual numbers $\mathbb{C}^0$ (both introduced by Clifford — see [22] for details). A 2-algebra $\mathbb{C}'$ is a commutative algebra with a basis $1, j$, where, unlike for $\mathbb{C}$-algebra, $j^2 = 1$. Therefore, this algebra is defined over reals and coincides the direct sum $\mathbb{R} \oplus \mathbb{R}$. This algebra contains nontrivial *idempotents*, i.e., the product of a number x multiplied by itself is the same number ($xx = x$). The second algebra considered here, a $\mathbb{C}^0$-algebra, is also a commutative 2-algebra with a basis $1, \epsilon$, where $\epsilon^2 = 0$ is a *nilpotent*. The nilpotent is an element that gives 0 during raising it to a positive integer power, i.e., $\epsilon^p = 0$ for $p \in \mathbb{Z}^+$. Both of these algebras contain zero divisors, which exclude them from consideration as candidates of a space for constructing fractal sets. The definition of sets in $\mathbb{C}'$ can be found in [84, 114], and their generalized version in [97]. The authors called the resulting sets obtained from the recursive equation of type (1.5) the *hyperbrot*, referring to the name of the Mandelbrot set. However, as the authors proved [97], the resulting set has the shape of a square, with a side length of $2\sqrt{2}/3\sqrt{3}$ (see Figure 2.16a), thus, it has no fractal properties. The sets defined in $\mathbb{C}^0$ are not connected, and, thus, cannot form $\mathcal{M}$- and $\mathcal{J}$-sets. An example of such a set of type (1.1), where $z, c \in \mathbb{C}^0$ is presented in Figure 2.16b.

The $\mathbb{H}$-algebra has more derivative associative 4-algebras. The first considered algebra is an algebra of split quaternions (or coquaternions) $\mathbb{H}'$, with a basis $1, i, j_1, j_2$, where $i^2 = -1$, $j_1^2 = j_2^2 = 1$. Similarly, as for $\mathbb{C}$-algebra, which has a derivative $\mathbb{C}^0$, the $\mathbb{H}$-algebra has its derivative $\mathbb{H}^0$, known as semiquaternions, with a basis $1, i, \kappa, \eta$, where $i^2 = -1$, $\kappa^2 = \eta^2 = 0$, $i\kappa = -\kappa i = \eta$, $\eta i = -i\eta = \kappa$, $\kappa\eta = \eta\kappa = 0$. The next algebra is an algebra of split semiquaternions $\mathbb{H}'^0$, with a basis $1, j, \kappa, \eta$, where $j^2 = 1$, $\kappa^2 = \eta^2 = 0$, $j\kappa = -\kappa j = \eta$, $\eta j = -j\eta = -\kappa$, $\kappa\eta = \eta\kappa = 0$. The last derivative of $\mathbb{H}$ is an algebra of $\frac{1}{4}$-quaternions $\mathbb{H}^{00}$, with a basis $1, \kappa, \eta, \omega$, where $\kappa^2 = \eta^2 = \omega^2 = 0$, $\kappa\eta = -\eta\kappa = \omega$, $\kappa\omega = \omega\kappa = \eta\omega = \omega\eta = 0$. These algebras contain zero divisors, nilpotents, and nontrivial idempotents, thus, cannot be considered as appropriate vector spaces for construction of fractal sets. Similarly, as in the case of split complex numbers, the recursive equation of type (1.5) produces regularly shaped geometrical structures that have no fractal properties. The examples of such sets constructed in $\mathbb{H}'$-space are presented in Figure 2.17. Next, the semiquaternionic sets can be constructed from the rules described in [87] (see Figure 2.18). Analogously, the set constructed in $\mathbb{H}'^0$-space can be visualized (see Figure 2.19). The details on algebraic operations on split semiquaternions can be found in [58].

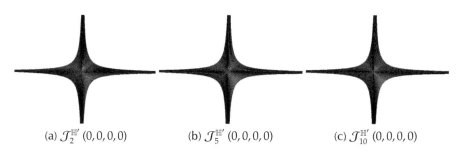

(a) $\mathcal{J}_2^{\mathbb{H}'}(0,0,0,0)$ (b) $\mathcal{J}_5^{\mathbb{H}'}(0,0,0,0)$ (c) $\mathcal{J}_{10}^{\mathbb{H}'}(0,0,0,0)$

Figure 2.17: Exemplary 3-D projections of sets in an $\mathbb{H}'$-space.

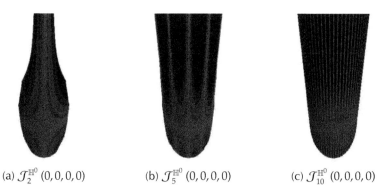

(a) $\mathcal{J}_2^{\mathbb{H}^0}(0,0,0,0)$ (b) $\mathcal{J}_5^{\mathbb{H}^0}(0,0,0,0)$ (c) $\mathcal{J}_{10}^{\mathbb{H}^0}(0,0,0,0)$

Figure 2.18: Exemplary 3-D projections of sets in an $\mathbb{H}^0$-space.

Finally, the sets constructed in $\mathbb{H}^{00}$-space do not give any resulting

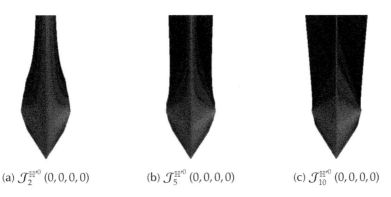

(a) $\mathcal{J}_2^{\mathbb{H}'^0} (0,0,0,0)$ (b) $\mathcal{J}_5^{\mathbb{H}'^0} (0,0,0,0)$ (c) $\mathcal{J}_{10}^{\mathbb{H}'^0} (0,0,0,0)$

Figure 2.19: Exemplary 3-D projections of sets in an $\mathbb{H}'^0$-space.

shape during visualization. The description on algebraic operations of $\frac{1}{4}$-quaternions can be found in [59].

Using the same transition processes, one can obtain derivatives of $\mathbb{O}$ — the alternative 8-algebras of split octonions $\mathbb{O}'$, semioctonions $\mathbb{O}^0$, split semioctonions $\mathbb{O}'^0$, $\frac{1}{4}$-octonions $\mathbb{O}^{00}$, split $\frac{1}{4}$-octonions $\mathbb{O}'^{00}$, and $\frac{1}{8}$-octonions $\mathbb{O}^{000}$. All of these algebras contain zero divisors. These transition processes are also applicable for the higher-dimensional algebras, but due to failing of conditions (1.2)–(1.4), they cannot generate the normed division algebras without zero divisors. Now, we can finalize the list of vector spaces appropriate for construction of $\mathcal{M}$- and $\mathcal{J}$-sets. However, there is another way to get new classes of vector spaces with desired properties: the already mentioned tensor product of algebras described in detail in the next chapter.

CHAPTER 3

Tensor Product Fractal Mutants

The tensor product operation on algebras mentioned in the previous chapter gives us a wide range of possibilities of construction of new vector spaces with unique properties. Due to the desired properties, like closing under addition and multiplication, some of them can be considered for a construction of $\mathcal{M}$- and $\mathcal{J}$-sets. In this chapter, we will take a closer look at rich possibilities given by a tensor product operation performed on algebras, and discuss in detail several vector spaces resulting from tensor product algebras, as well as $\mathcal{M}$- and $\mathcal{J}$-sets constructed in them. Since such vector spaces are composed from at least two elementary vector spaces, we can facetiously call the fractal structures generated in these spaces fractal mutants.

3.1 BICOMPLEX, TRICOMPLEX, ... MULTICOMPLEX FRACTAL SETS

3.1.1 Julia and Mandelbrot Sets in Bicomplex Vector Space

Considering the definition of a tensor product presented in Section 2.3.1, one can obtain the simplest hypercomplex tensor product algebra $\mathbb{C} \otimes \mathbb{C}$, the algebra of bicomplex numbers (known also as tessarines) proposed by James Cockle in a series of papers [24, 25, 26, 27]. It is a 4-algebra with a basis of $1, i_1, i_2, j$, where $i_1^2 = i_2^2 = -1$, $j^2 = 1$, $i_1 i_2 = i_2 i_1 = j$. Here, we can introduce the following notation of bicomplex numbers: $\mathbb{C}_2$, which is equivalent to $\mathbb{C} \otimes \mathbb{C}$. This algebra has properties similar to $\mathbb{H}$: Both algebras are four-dimensional and have a symbolic form given by (2.2), and both

TABLE 3.1
Multiplication Table for
Bicomplex Numbers.

×	1	i_1	i_2	j
1	1	i_1	i_2	j
i_1	i_1	-1	j	$-i_2$
i_2	i_2	j	1	i_1
j	j	$-i_2$	i_1	-1

are associative and alternative; however, the difference is in the type of imaginary units (cf. basis of $\mathbb{C}_2$ and $\mathbb{H}$) and the commutativity property of $\mathbb{C}_2$. Due to the existence of idempotents for $\mathbb{C}_2$,

$$z_1 + z_2 i_2 = (z_1 - z_2 i_1)e_1 + (z_1 + z_2 i_1)e_2, \tag{3.1}$$

where $z_1, z_2 \in \mathbb{C}_1 := \{x + yi_1 | i_1^2 = -1\}$ (or just $\mathbb{C}$, since $\mathbb{C} \equiv \mathbb{C}_1$), $e_1 = (1 + j)/2$, $e_2 = (1 - j)/2$, the addition, multiplication, and division operations become much easier, since they can be performed element-wise. Considering that $z_1 = z_{11} + iz_{12}$, $z_2 = z_{21} + iz_{22}$, $z_m \in \mathbb{C}_2$, and $z_{mn} \in \mathbb{C}_1$, the addition of two bicomplex numbers is very similar to the addition of quaternions given by (2.3) (see [78] for details):

$$z_1 + z_2 := (z_{11} + z_{21}) + i(z_{12} + z_{22}), \tag{3.2}$$

while the multiplication is given by:

$$z_1 \cdot z_2 := (z_{11} + iz_{12})(z_{21} + iz_{22}) = (z_{11}z_{21} - z_{12}z_{22}) + i(z_{11}z_{22} + z_{21}z_{12}), \tag{3.3}$$

and is ruled by the multiplication table (see Table 3.1) of bicomplex numbers (see Table 2.1).

Due to the mentioned similarity of $\mathbb{C}_2$ to $\mathbb{H}$, one could expect subsequent similarities in construction, properties, and visualization of $\mathcal{M}$- and $\mathcal{J}$-sets in $\mathbb{C}_2$. Indeed, these spaces and fractal sets have a lot of common properties. But let us start from the formal definition of the generalized $\mathcal{J}_p^{\mathbb{C}_2}$-sets [107]:

$$\mathcal{J}_p^{\mathbb{C}_2} = \left\{c \in \mathbb{C}_2 | f^{(s)}(z) \nrightarrow \infty \text{ as } s \to \infty\right\}, \text{ for } p \geq 2, \tag{3.4}$$

and, subsequently, the generalized $\mathcal{M}_p^{\mathbb{C}_2}$-set

$$\mathcal{M}_p^{\mathbb{C}_2} = \left\{c \in \mathbb{C}_2 | f^{(s)}(0) \nrightarrow \infty \text{ as } s \to \infty\right\}, \text{ for } p \geq 2, \tag{3.5}$$

where $f^{(s)}(\cdot)$ denotes a function composition performed s times.

Similarly to $\mathcal{J}^{\mathbb{C}}$-, $\mathcal{M}^{\mathbb{C}}$-, $\mathcal{J}^{\mathbb{H}}$-, and $\mathcal{M}^{\mathbb{H}}$-sets, the bailout value for $\mathcal{J}^{\mathbb{C}_2}$- and $\mathcal{M}^{\mathbb{C}_2}$-sets equals 2 [107]. Moreover, as it was proven by Rochon [108], the $\mathcal{M}_p^{\mathbb{C}_2}$-set is connected, while for $\mathcal{J}_p^{\mathbb{C}_2}$-sets, in contrast to $\mathcal{J}_p^{\mathbb{C}}$-sets, there are the following three cases:

1. For $c \in \mathcal{M}_p^{\mathbb{C}_2}$, the $\mathcal{J}_p^{\mathbb{C}_2}$-sets are connected.

2. For $c \notin \mathcal{M}_p^{\mathbb{C}_2}$ and $c \in SA^{\mathbb{C}_2}(\infty)$, the $\mathcal{J}_p^{\mathbb{C}_2}$-sets are totally disconnected (i.e., homeomorphic to the Cantor dust).

3. For all other cases of c, the $\mathcal{J}_p^{\mathbb{C}_2}$-sets are disconnected, but not totally.

Note that $SA^{\mathbb{C}_2}(\infty)$ is a *strong basin of attraction* at infinity:

$$SA^{\mathbb{C}_2}(\infty) = A_{c_1 - c_2 i_1}(\infty) \times A_{c_1 + c_2 i_1}(\infty), \tag{3.6}$$

for $z \to z^2 + c$, where $c \in \mathbb{C}_2$ is given by (3.1), following the definition of Rochon [108].

The visualization techniques used for $\mathcal{J}^{\mathbb{C}_2}$- and $\mathcal{M}^{\mathbb{C}_2}$-sets (or *bicomplex filled Julia sets* and the *Tetrabrot*, respectively, following the nomenclature introduced by Rochon [107, 108]) are quite similar to quaternionic fractal sets, since both types of fractal sets are represented in a 4-space. An interesting approach of bicomplex distance estimation and ray-tracing techniques for $\mathcal{J}^{\mathbb{C}_2}$- and $\mathcal{M}^{\mathbb{C}_2}$-sets was proposed in [81]. The exemplary $\mathcal{J}^{\mathbb{C}_2}$-sets are presented in Figure 3.1.

Similarly to the case of $\mathcal{J}^{\mathbb{H}}$-sets, the $\mathcal{J}^{\mathbb{C}_2}$-sets are trivial when in the recurrence equation of a type (1.1), the constant $c \in \mathbb{C}$ instead of $c \in \mathbb{C}_2$, i.e., when the last two coefficients of the $c \in \mathbb{C}_2$ equal zero. This triviality is even better observed in the case of bicomplex fractal sets, since a 3-D projection of a quaternionic trivial fractal set is obtained from the rotation of a fractal set in a $\mathbb{C}$-plane around the axis of reals, while in the case of trivial bicomplex fractal sets, their 3-D projections are obtained from extrusion of a fractal set in a $\mathbb{C}$-plane (see Figure 3.2 and Figure 2.2). Several other examples of trivial $\mathcal{J}^{\mathbb{C}_2}$-sets, similar to previously analyzed $\mathcal{J}^{\mathbb{H}}$-sets (presented in Figure 2.2) with respect to the value of c, are presented in Figure 3.3.

Now, let us consider a case of $\mathcal{J}_p^{\mathbb{C}_2}$-sets with various p values. The recursive equation considered here is of a type of (1.5). Analogously to the quaternionic fractal sets, a few bicomplex fractal sets with various p

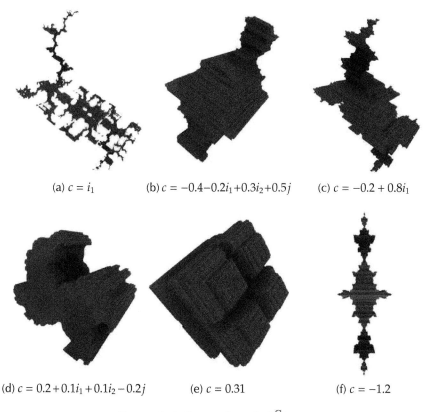

(a) $c = i_1$ (b) $c = -0.4-0.2i_1+0.3i_2+0.5j$ (c) $c = -0.2 + 0.8i_1$

(d) $c = 0.2+0.1i_1+0.1i_2-0.2j$ (e) $c = 0.31$ (f) $c = -1.2$

Figure 3.1: Examples of $\mathcal{J}_2^{\mathbb{C}^2}$-sets.

values were generated and presented in Figure 3.4. Surprisingly, a shape of resulting 3-D projection of a fractal set is not the $(n-1)$-sphere, but is the *Steinmetz solid* (see Figure 3.4f), known also as a *bicylinder*. This solid results from the intersection of two cylinders at right angles. More information on this solid and its properties can be found in [86]. One can observe in Figure 3.4 that the convergence of the tending of $\mathcal{J}_p^{\mathbb{C}^2}$-sets to the Steinmetz solid is quicker than in the case of quaternionic analogue (see Figure 2.7). The results of studies on the convergence of $\mathcal{M}_p^{\mathbb{C}^2}$- and $\mathcal{J}_p^{\mathbb{C}^2}$-sets can be found in [65].

Considering that visual representation conditions of bicomplex fractal sets are identical to those of quaternionic fractal sets, since both types of 4-space fractals are mapped into the 3-D Euclidean space, the relations (2.5)-(2.8) are also true for bicomplex fractals. The method of setting the last element of a bicomplex number to zero ($j = 0$) is also applicable for

(a)   (b)

Figure 3.2: 3-D projection of $\mathcal{J}^{\mathbb{C}_2}(-0.5, 0.5, 0, 0)$ and its cross-section along the $\mathbb{C}$-plane.

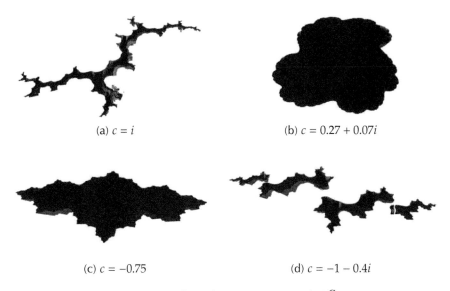

(a) $c = i$ (b) $c = 0.27 + 0.07i$

(c) $c = -0.75$ (d) $c = -1 - 0.4i$

Figure 3.3: Examples of cross-sectioned $\mathcal{J}^{\mathbb{C}_2}$-sets.

visual representation of the bicomplex fractal sets that where used by Rochon [107]. In order to present what a bicomplex fractal set looks like, the sequences of 3-D slices with a discrete step were generated. These slices obtained while moving the cutting hyperplane along the axis of reals and i_1-axis are presented in Figure 3.5. The slices obtained while moving the cutting hyperplane along i_1-axis and i_2-axis are presented in Figure 3.6.

The $\mathcal{M}_p^{\mathbb{C}_2}$-set defined by (3.5) is a kind of generalization of $\mathcal{J}_p^{\mathbb{C}_2}$-sets,

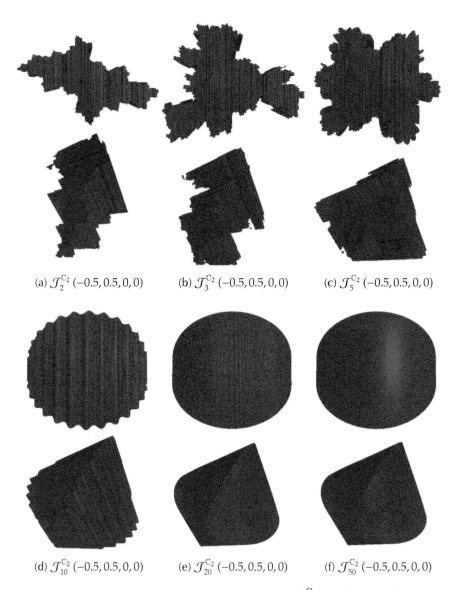

(a) $\mathcal{J}_2^{\mathbb{C}_2}(-0.5, 0.5, 0, 0)$ (b) $\mathcal{J}_3^{\mathbb{C}_2}(-0.5, 0.5, 0, 0)$ (c) $\mathcal{J}_5^{\mathbb{C}_2}(-0.5, 0.5, 0, 0)$

(d) $\mathcal{J}_{10}^{\mathbb{C}_2}(-0.5, 0.5, 0, 0)$ (e) $\mathcal{J}_{20}^{\mathbb{C}_2}(-0.5, 0.5, 0, 0)$ (f) $\mathcal{J}_{50}^{\mathbb{C}_2}(-0.5, 0.5, 0, 0)$

Figure 3.4: Tending of 3-D projections of $\mathcal{J}_p^{\mathbb{C}_2}(-0.5, 0.5, 0, 0)$ to the Steinmetz solid.

since, as Rochon stated in [107], there exists a dichotomy between the above-mentioned sets, similar to their analogues on a $\mathbb{C}$-plane. The $\mathcal{M}_2^{\mathbb{C}_2}$-

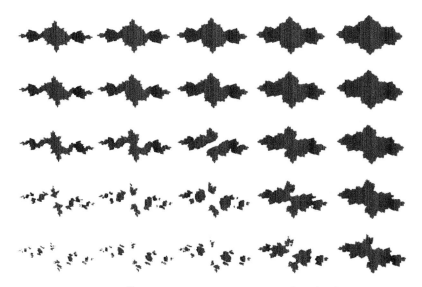

Figure 3.5: Slices of $\mathcal{J}^{\mathbb{C}_2}(-1, 0.1, 0, 0)$ along axis of reals (from -1 to -0.6) and i_1-axis (from 0.1 to 0.5), with a step of 0.1.

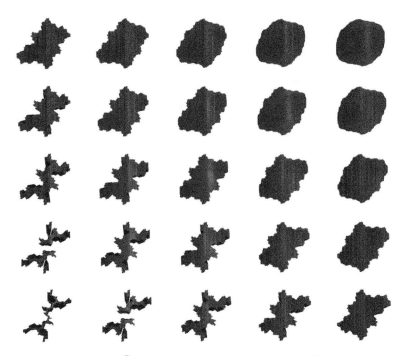

Figure 3.6: Slices of $\mathcal{J}^{\mathbb{C}_2}(0, -0.5, 0.1, 0)$ along i_1-axis (from -0.5 to -0.1) and i_2-axis (from 0.1 to 0.5), with a step of 0.1.

set is presented in Figure 3.7. Analogously to $\mathcal{J}_p^{\mathbb{C}_2}$-sets, when $p \to \infty$, the 3-D projection of a shape of the $\mathcal{M}_p^{\mathbb{C}_2}$-set tends to the Steinmetz solid.

Figure 3.7: 3-D projection of the $\mathcal{M}_2^{\mathbb{C}_2}$-set (the Tetrabrot).

Several studies on generalizations of the $\mathcal{M}_p^{\mathbb{C}_2}$-set for $p \in \mathbb{Z}^+$ were performed by Wang and Song [118]. They studied the dynamics of the $\mathcal{M}_p^{\mathbb{C}_2}$-set and proved that considering the bicomplex map of type (1.1) for $z, c \in \mathbb{C}_2$ and $p > 1$, the critical point of $f(z)$ is zero, and for $p \in [0, 1]$, the resulting set is a degenerated $\mathcal{M}_p^{\mathbb{C}_2}$-set. They also showed particular cases: When $p = 1$, the critical point does not exist for $f(z)$, and for $0 \leq p < 1$, the critical point is at the infinity. Several examples of the $\mathcal{M}_p^{\mathbb{C}_2}$-set of higher order polynomials are available in [118].

3.1.2 Julia and Mandelbrot Sets in Tricomplex Vector Space

Developing the idea of tensor product algebras, the next hypercomplex tensor product algebra after $\mathbb{C}_2$ is a tricomplex algebra $\mathbb{C} \otimes \mathbb{C} \otimes \mathbb{C}$, or simply $\mathbb{C}_3$. This is an 8-algebra introduced by Corrado Serge with a basis $1, i_1, i_2, j_1, i_3, j_2, j_3, i_4$, where $i_1^2 = i_4^2 = -1$, $i_4 = i_1 j_3 = i_1 i_2 i_3$, $j_2 = i_1 i_3 = i_3 i_1$, $j_2^2 = 1$, $j_1 = i_1 i_2 = i_2 i_1$, and $j_1^2 = 1$ [97]. This 8-algebra is similar in some ways to the algebra of octonions, however, a $\mathbb{C}_3$-algebra is commutative, in contrast to an $\mathbb{O}$-algebra, which is alternative only (compared with the properties described in Section 2.2.1). The set of tricomplex numbers can be presented using the following equivalent representations [97]:

$$\mathbb{C}_3 := \left\{ \eta = \varsigma_1 + \varsigma_2 i_3 \,\middle|\, \varsigma_1, \varsigma_2 \in \mathbb{C}_2 \right\} = \left\{ z_1 + z_2 i_2 + z_3 i_3 + z_4 j_3 \,\middle|\, z_{1,\ldots,4} \in \mathbb{C} \right\}$$
$$= \left\{ a_1 + a_2 i_1 + a_3 i_2 + a_4 i_3 + a_5 i_4 + a_6 j_1 + a_7 j_2 + a_8 j_3 \,\middle|\, a_{1,\ldots,8} \in \mathbb{R} \right\}. \quad (3.7)$$

Due to the existence of idempotents in $\mathbb{C}_3$, which can be represented in terms of two bicomplex elements [97],

$$\eta = (\varsigma_1 - \varsigma_2 i_2)\,\gamma_2 + (\varsigma_1 + \varsigma_2 i_2)\,\bar{\gamma}_2, \quad \varsigma_1, \varsigma_2 \in \mathbb{C}_2, \tag{3.8}$$

where $\gamma_2 = (1 + j_3)/2$ and $\bar{\gamma}_2 = (1 - j_3)/2$, or, in terms of four complex elements [97],

$$\eta = w_1 \gamma_1 \gamma_2 + w_2 \gamma_1 \bar{\gamma}_2 + w_3 \bar{\gamma}_1 \gamma_2 + w_4 \bar{\gamma}_1 \bar{\gamma}_2, \quad w_{1,\dots,4} \in \mathbb{C} \tag{3.9}$$

where $w_1 = (z_1 + z_4) - (z_2 - z_3)\,i_1$, $w_2 = (z_1 + z_4) + (z_2 - z_3)\,i_1$, $w_3 = (z_1 - z_4) - (z_2 + z_3)\,i_1$, and $w_4 = (z_1 - z_4) + (z_2 + z_3)\,i_1$, the addition and multiplication of tricomplex numbers can be performed element-wise. These operations on two tricomplex numbers η_1 and η_2 (which are necessary for construction of $\mathcal{J}^{\mathbb{C}_3}$- and $\mathcal{M}^{\mathbb{C}_3}$-sets) are defined as follows [97]:

$$\eta_1 + \eta_2 := (\varsigma_1 + \varsigma_3) + (\varsigma_2 + \varsigma_4)\,i_3, \tag{3.10}$$

$$\eta_1 \cdot \eta_2 := (\varsigma_1 \varsigma_3 - \varsigma_2 \varsigma_4) + (\varsigma_1 \varsigma_4 + \varsigma_2 \varsigma_3)\,i_3. \tag{3.11}$$

The multiplication is ruled by the special multiplication table (see Table 3.2) of tricomplex numbers (compared to Table 2.2).

TABLE 3.2　Multiplication Table for Tricomplex Numbers.

$\times$	1	i_1	i_2	i_3	i_4	j_1	j_2	j_3
1	1	i_1	i_2	i_3	i_4	j_1	j_2	j_3
i_1	i_1	-1	j_1	j_2	$-j_3$	$-i_2$	$-i_3$	i_4
i_2	i_2	j_1	-1	j_4	$-j_2$	$-i_1$	i_4	$-i_3$
i_3	i_3	j_2	j_3	-1	$-j_1$	i_4	$-i_1$	$-i_2$
i_4	i_4	$-j_3$	$-j_2$	$-j_1$	-1	i_3	i_2	i_1
j_1	j_1	$-i_2$	$-i_1$	i_4	i_3	1	$-j_3$	$-j_2$
j_2	j_2	$-i_3$	i_4	$-i_1$	i_2	$-j_3$	1	$-j_1$
j_3	j_3	i_4	$-i_3$	$-i_2$	i_1	$-j_2$	$-j_1$	1

Since tricomplex numbers are closed under addition and multiplication, the $\mathcal{J}^{\mathbb{C}_3}$- and $\mathcal{M}^{\mathbb{C}_3}$-sets can be defined in the same way as previously, i.e., replacing $\mathbb{C}_2$ by $\mathbb{C}_3$ in (3.4) and (3.5). As the authors of [97] proved, the $\mathcal{M}_p^{\mathbb{C}_3}$-sets are connected for $p \geq 2$. Considering also the generalized Fatou–Julia theorem [39] one can state the existence of three types of connectedness of $\mathcal{J}_p^{\mathbb{C}_3}$-sets, similar to $\mathcal{J}_p^{\mathbb{C}_2}$-sets (see Section 3.1.1 for details). The extended analysis presented in [39] shows that c values corresponding to connected $\mathcal{J}^{\mathbb{C}_3}$-sets are inside the $\mathcal{M}^{\mathbb{C}_3}$-set, while the other two types of $\mathcal{J}^{\mathbb{C}_3}$-sets (totally disconnected, and disconnected, but not totally) are out of the $\mathcal{M}^{\mathbb{C}_3}$-set.

The visualization of $\mathcal{J}^{\mathbb{C}_3}$- and $\mathcal{M}^{\mathbb{C}_3}$-sets seems to be problematic due to the same reasons as for octonionic sets, i.e., the mentioned sets exist in 8-D vector spaces, while only three-dimensional combinations of 8-D elements can be presented. This results in most information about the geometry of these objects to be lost. Nevertheless, Parisé and Rochon performed extended geometrical analysis of $\mathcal{J}^{\mathbb{C}_3}$- and $\mathcal{M}^{\mathbb{C}_3}$-sets in their paper [97]. The approach applied by the authors in [39, 97] is based on selection of particular triplets from the eight available elements of a tricomplex number and visualisation a 3-D slice of a given set. In the case of $\mathcal{J}^{\mathbb{C}_3}$- and $\mathcal{M}^{\mathbb{C}_3}$-sets, there are 56 possible combinations of such triplets [97]. The authors studied symmetries and relationships between these 3-D slices, and defined four principal types of such slices, which are related to imaginary units selected for a given triplet. Examples of these slices can be found in [97].

3.1.3   Julia and Mandelbrot Sets in Multicomplex Vector Space

The previously discussed bicomplex and tricomplex algebras can be generalized to a multicomplex algebra defined by n-tensor product of complex vector spaces $\mathbb{C} \otimes \mathbb{C} \otimes \ldots \otimes \mathbb{C}$ and denoted by $\mathbb{C}_n$. The symbolic form of a set of multicomplex numbers is given by [103]:

$$\mathbb{C}_n := \left\{ \xi_n = \xi_{n-1,1} + \xi_{n-1,2} i_n \,\middle|\, \xi_{n-1,1}, \xi_{n-1,2} \in \mathbb{C}_{n-1} \right\}, \qquad (3.12)$$

where $i_n^2 = -1$. The symbolic representation (3.12) can be decomposed to the form [39]

$$\mathbb{C}_n := \left\{ \xi_n = \xi_{n-2,1} + \xi_{n-2,2} i_{n-1} + \xi_{n-2,3} i_n \xi_{n-2,4} j_n \,\middle|\, \xi_{n-2,1,\ldots,4} \in \mathbb{C}_{n-2} \right\}, \qquad (3.13)$$

where $j_n = i_n i_{n-1} = i_{n-1} i_n$, $j_n^2 = 1$. Thus, every multicomplex number in $\mathbb{C}_n$ contains 2^n elements, with the associators defined in $\mathbb{R}$, or equivalently, 2^{n-m} elements defined in $\mathbb{C}_m$ for $0 \leq m \leq n$.

Similarly to (3.1), (3.8), and (3.9), it is possible to generalize an idempotent representation of (3.12) as follows [39]:

$$\xi_n = \left(\xi_{n-1,1} - \xi_{n-1,2} i_{n-1} \right) \gamma_{n-1} + \left(\xi_{n-1,1} + \xi_{n-1,2} i_{n-1} \right) \bar{\gamma}_{n-1}, \qquad (3.14)$$

where

$$\gamma_{n-1} := \frac{1 + i_{n-1} i_n}{2} = \frac{1 + j_n}{2}, \quad \bar{\gamma}_{n-1} := \frac{1 - i_{n-1} i_n}{2} = \frac{1 - j_n}{2}. \qquad (3.15)$$

The idempotents have the following properties [39]:

$$\gamma_{n-1}^2 = \gamma_{n-1}, \ \bar{\gamma}_{n-1}^2 = \bar{\gamma}_{n-1}, \ \gamma_{n-1} + \bar{\gamma}_{n-1} = 1, \ \gamma_{n-1}\bar{\gamma}_{n-1} = \bar{\gamma}_{n-1}\gamma_{n-1} = 0.$$
(3.16)

Considering (3.14)–(3.16), one can state that $\mathbb{C}_n$ is commutative and associative. Due to these properties, the multicomplex numbers are closed under addition and multiplication for $n \in \mathbb{Z}^+$, which means that addition and multiplication of two multicomplex numbers, $\xi_{n,1}$ and $\xi_{n,2}$, can be performed element-wise:

$$\xi_{n,1} + \xi_{n,2} := (\xi_{n-1,1} + \xi_{n-1,3}) + (\xi_{n-1,2} + \xi_{n-1,4})\, i_n,$$
(3.17)

$$\xi_{n,1} \cdot \xi_{n,1} := (\xi_{n-1,1}\xi_{n-1,3} - \xi_{n-1,2}\xi_{n-1,4}) + (\xi_{n-1,1}\xi_{n-1,4} + \xi_{n-1,2}\xi_{n-1,3})\, i_n.$$
(3.18)

These properties lead to a possibility of generalizing fractal sets to the form of $\mathcal{J}^{\mathbb{C}_n}$- and $\mathcal{M}^{\mathbb{C}_n}$-sets,

$$\mathcal{J}_p^{\mathbb{C}_n} = \left\{ c \in \mathbb{C}_n \mid f^{(s)}(z) \nrightarrow \infty \text{ as } s \to \infty \right\}, \text{ for } p \geq 2,$$
(3.19)

and, subsequently, the generalized $\mathcal{M}_p^{\mathbb{C}_2}$-set,

$$\mathcal{M}_p^{\mathbb{C}_n} = \left\{ c \in \mathbb{C}_n \mid f^{(s)}(0) \nrightarrow \infty \text{ as } s \to \infty \right\}, \text{ for } p \geq 2,$$
(3.20)

where $f^{(s)}(\cdot)$ denotes a function composition performed s times.

As the authors of [39] proved, the $\mathcal{M}_p^{\mathbb{C}_n}$-sets are connected, and, following the generalized Fatou–Julia theorem presented in the same paper, the $\mathcal{J}_p^{\mathbb{C}_n}$-sets reveal three types of connectedness (see Section 3.1.1 for details).

3.2 TENSOR PRODUCT HYPERCOMPLEX AND MULTIHYPERCOMPLEX FRACTAL SETS

3.2.1 Julia and Mandelbrot Sets in Biquaternionic Vector Space

Obviously, one can construct a great variety of tensor product algebras that create vector spaces with appropriate conditions for the existence of fractal sets. One such tensor product algebras interesting to us due to its properties is the algebra of *biquaternions* $\mathbb{C} \otimes \mathbb{H}$, known also as an algebra of *complex quaternions* or *complexified quaternions* $\tilde{q}$, since any $\tilde{q} \in \mathbb{C} \otimes \mathbb{H}$ can be presented as [68]

$$\tilde{q} = \mathfrak{R}(\tilde{q}) + i\mathfrak{I}(\tilde{q}),$$
(3.21)

TABLE 3.3
Multiplication Table for
Biquaternions for the
Representation (3.22).

×	1	i_1	i_2	i_3
1	1	i_1	i_2	i_3
i_1	i_1	-1	$-i_2$	$-i_3$
i_2	i_2	$-i_3$	1	i_1
i_3	i_3	$-i_2$	i_1	1

where $\Re(\tilde{q}) = \sum_{n=0}^{3} \Re(\tilde{q}_n i_n)$ and $\Im(\tilde{q}) = \sum_{n=0}^{3} \Im(\tilde{q}_n i_n)$. This is a 4-algebra with a basis $1, i_1, i_2, i_3$ with a symbolic representation

$$\mathbb{C} \otimes \mathbb{H} := \left\{ \tilde{q} = a_1 + a_2 i_1 + a_3 i_2 + a_4 i_3 \middle| a_n \in \mathbb{C} \right\}, \tag{3.22}$$

proposed by Hamilton in 1844 as an alternative form of the previously developed "real" quaternions. The symbolic representation (3.22) can be also presented in the alternative form

$$\mathbb{C} \otimes \mathbb{H} := \{ \tilde{q} = (b_1 + d_1 j) + (b_2 + d_2 j) i_1 + (b_3 + d_3 j) i_2$$
$$+ (b_4 + d_4 j) i_3 \middle| b_n, d_n \in \mathbb{R} \}, \tag{3.23}$$

where $j^2 = -1$, $i_1^2 = i_2^2 = i_3^2 = i_1 i_2 i_3 = -1$.

The biquaternions retain the properties of quaternions, i.e., they are associative and alternative, thus, the operation of addition and multiplication of biquaternions can be defined in the same way as for quaternions, and are given by (2.3) and (2.4), respectively. Depending on the chosen symbolic representation (given by (3.22) and (3.23)), the multiplication tables can be presented as follows (Tables 3.3 and 3.4).

The $\mathcal{M}^{\mathbb{C} \otimes \mathbb{H}}$- and $\mathcal{J}^{\mathbb{C} \otimes \mathbb{H}}$-sets can be defined as their quaternionic and bicomplex analogues, similarly to, e.g., (3.4) and (3.5). Both of these sets are bounded by the definition. This introduces the generalized $\mathcal{M}$-$\mathcal{J}$ sets in terms of biquaternions, with dynamics similar to the previously presented $\mathcal{M}^{\mathbb{H}}$-, $\mathcal{J}^{\mathbb{H}}$-, $\mathcal{M}^{\mathbb{C}_2}$-, and $\mathcal{J}^{\mathbb{C}_2}$-sets. Finally, the $\mathcal{M}_p^{\mathbb{C} \otimes \mathbb{H}}$-set can be alternatively presented for a certain bailout value B as follows:

$$\mathcal{M}_p^{\mathbb{C} \otimes \mathbb{H}} = \left\{ c \in \mathbb{C} \otimes \mathbb{H} \middle| f_c^{(s)}(0) \leq B \ \forall s \in \mathbb{N} \right\} \text{ for } p \geq 2, \tag{3.24}$$

where $f_c^{(s)}(\cdot)$ denotes a function composition performed s times.

Similarly, as in the previous cases, it is suitable to limit $\mathcal{M}_p^{\mathbb{C} \otimes \mathbb{H}}$- and $\mathcal{J}_p^{\mathbb{C} \otimes \mathbb{H}}$-sets by a 3-sphere with a radius of 2.

TABLE 3.4 Multiplication Table for Biquaternions for the Representation (3.23).

$\times$	1	j	i_1	ji_1	i_2	ji_2	i_3	ji_3
1	1	j	i_1	ji_1	i_2	ji_2	i_3	ji_3
j	j	1	$-i_3$	ji_2	$-ji_3$	ji_1	$-i_1$	$-i_2$
i_1	i_1	i_3	1	$-i_2$	$-ji_1$	$-ji_3$	j	$-ji_2$
ji_1	ji_1	$-ji_2$	i_2	1	i_1	$-j$	$-ji_3$	$-i_3$
i_2	i_2	$-ji_3$	ji_1	$-i_1$	-1	i_3	$-ji_2$	j
ji_2	ji_2	$-ji_1$	$-ji_3$	j	$-i_3$	-1	i_2	i_1
i_3	i_3	i_1	$-j$	$-ji_3$	ji_2	$-i_2$	-1	ji_1
ji_3	ji_3	$-i_2$	$-ji_2$	$-i_3$	j	i_1	ji_1	-1

Only a few mentions on biquaternionic fractal sets can be found in the available literature. Gintz [41] introduced these fractal sets and presented preliminaries on them and several of their 3-D cross-sections, as well as performed their graphical analysis. A year later, the authors of [12] studied symmetry properties of biquaternionic $\mathcal{J}$-sets. The mathematical fundamentals on biquaternionic fractal sets with appropriate graphical analysis were introduced in [64].

The connectedness of $\mathcal{M}_p^{\mathbb{C}\otimes\mathbb{H}}$- and $\mathcal{J}_p^{\mathbb{C}\otimes\mathbb{H}}$-sets is similar to those of $\mathcal{J}^{\mathbb{C}_2}$- and $\mathcal{M}^{\mathbb{C}_2}$-sets (see Section 3.1.1), i.e., $\mathcal{M}_p^{\mathbb{C}\otimes\mathbb{H}}$-sets are connected, while $\mathcal{J}_p^{\mathbb{C}\otimes\mathbb{H}}$-sets can be connected, totally disconnected, and disconnected, but not totally. The examples of the last type of connectedness are shown in Figure 3.8.

Analyzing the geometrical structure of $\mathcal{J}^{\mathbb{C}\otimes\mathbb{H}}$-sets, one can observe similarities in the shapes of resulting fractal sets comparable with $\mathcal{J}^{\mathbb{C}_2}$-sets and $\mathcal{J}^{\mathbb{H}}$-sets. This is due to the similarity of construction of algebraic formulas between these sets. Similarly, as in the case of bicomplex $\mathcal{J}$-sets, by cutting the $\mathcal{J}^{\mathbb{C}\otimes\mathbb{H}}$-sets along the axis of reals, one can obtain the well-known $\mathcal{J}$-sets on a $\mathbb{C}$-plane. The biquaternionic analogues of $\mathcal{J}$-sets are presented in Figure 3.9. One can observe similarities with $\mathcal{J}$-sets on a $\mathbb{C}$-plane (compare to Figure 1.3), as well as with quaternionic (compare to Figure 2.3) and bicomplex (compare to Figure 3.3) fractal sets. When analysing the shapes of 3-D cross-sections of biquaternionic $\mathcal{J}$-sets one can observe that their symmetry properties are broken with respect to quaternionic analogues of these $\mathcal{J}$-sets. This is very well visible in Figures 3.9b and 3.9f which are neither as rotationally symmetric as their quaternionic analogues, nor as quadrilaterally symmetric as their bicomplex analogues, which results from the complexification of

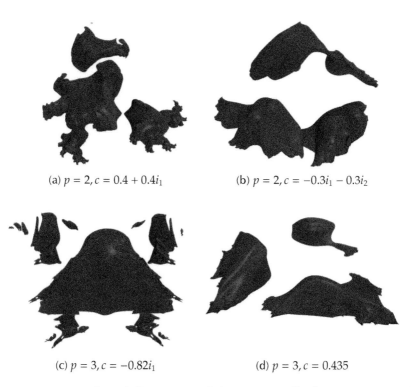

(a) $p = 2, c = 0.4 + 0.4i_1$

(b) $p = 2, c = -0.3i_1 - 0.3i_2$

(c) $p = 3, c = -0.82i_1$

(d) $p = 3, c = 0.435$

Figure 3.8: Examples of disconnected, but not totally, biquaternionic $\mathcal{J}$-sets.

a quaternion. For visual representation of the shape variability of the biquaternionic $\mathcal{J}$-set, the sets of 3-D cross-sections are presented in Figure 3.11, and in Figure 3.12, while moving the cutting hyperplane along i_1-axis and i_2-axis, respectively.

Considering the generalized version of biquaternionic $\mathcal{J}$-sets, $\mathcal{J}_p^{\mathbb{C}\otimes\mathbb{H}}$, with various p values, described by the recursive equation of type (1.5), one can observe several interesting properties. First of all, the shape of $\mathcal{J}_p^{\mathbb{C}\otimes\mathbb{H}}$-sets for $c = 0$ does not tend to regular shapes, as in the case of $\mathcal{J}_p^{\mathbb{C}}$-sets (see Figure 1.7), $\mathcal{J}_p^{\mathbb{H}}$-sets (see Figure 2.7), or $\mathcal{J}_p^{\mathbb{C}^2}$-sets (see Figure 3.4). The resulting shapes remain irregular, and, surprisingly, they also remain fractal (see Figure 3.10).

In the case of a biquaternionic quadratic map, the resulting set (see Figure 3.10a) is not a 3-sphere (like in the case of $\mathcal{J}_p^{\mathbb{H}}$-sets) and is not the 4-dimensional Steinmetz solid (like in the case of $\mathcal{J}_p^{\mathbb{C}^2}$-sets) due to

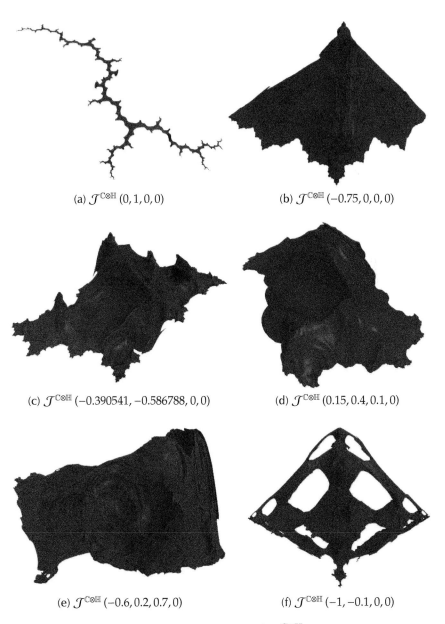

(a) $\mathcal{J}^{\mathbb{C}\otimes\mathbb{H}}(0,1,0,0)$

(b) $\mathcal{J}^{\mathbb{C}\otimes\mathbb{H}}(-0.75,0,0,0)$

(c) $\mathcal{J}^{\mathbb{C}\otimes\mathbb{H}}(-0.390541,-0.586788,0,0)$

(d) $\mathcal{J}^{\mathbb{C}\otimes\mathbb{H}}(0.15,0.4,0.1,0)$

(e) $\mathcal{J}^{\mathbb{C}\otimes\mathbb{H}}(-0.6,0.2,0.7,0)$

(f) $\mathcal{J}^{\mathbb{C}\otimes\mathbb{H}}(-1,-0.1,0,0)$

Figure 3.9: Examples of $\mathcal{J}^{\mathbb{C}\otimes\mathbb{H}}$-sets.

the tensor product of dimensionally unequal algebras. When the order of an iterated polynomial increases, the resulting $\mathcal{J}$-sets resemble a shape of a pillow with multiple horn-like protruding geometric structures on

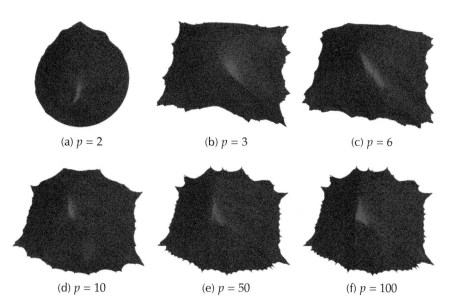

(a) $p = 2$ (b) $p = 3$ (c) $p = 6$

(d) $p = 10$ (e) $p = 50$ (f) $p = 100$

Figure 3.10: Examples of $\mathcal{J}_p^{\mathbb{C} \otimes \mathbb{H}}$-sets for various values of p and with $c = 0$.

the boundaries. Therefore, these sets can be introduced with the name of the *Devil's pillows*. The Devil's pillows are not rotationally symmetric; however, they retain several symmetry planes.

3.2.2 Beyond Biquaternionic Julia and Mandelbrot Sets

Increasing the dimensionality of the multiplied algebras (that form tensor product algebras) used for the construction of hypercomplex fractal sets, the number of possibilities also increases; however, there are several limitations, e.g., resulted from the conditions (1.2)–(1.4). Thus, the number of possibilities is limited. The other higher-dimensional tensor product algebras are described much less than the previously analyzed ones, while the fractal sets that can (or cannot) be constructed using them are not described at all in the literature. Due to the high dimensionality of these algebras (which starts from 8), it is difficult to construct fractal sets of the $\mathcal{M}$- and $\mathcal{J}$-type, as well as investigate their properties. However, let us try to analyze such tensor product algebras and perform at least their preliminary selection.

Starting from tensor products of algebras of quaternions, let us first consider the quaterquaternions. The 8-algebra of quaterquaternions $\mathbb{H}_2$ contains zero-divisors [2], however, there is a possibility

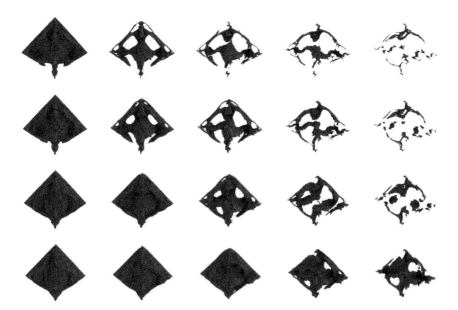

Figure 3.11: Slices of $\mathcal{J}^{\mathbb{C}\otimes\mathbb{H}_2}(-1,0,0,0)$ along the axis of reals (from -1 to -0.7) and i_1-axis (from 0 to 0.4), with a step of 0.1.

Figure 3.12: Slices of $\mathcal{J}^{\mathbb{C}\otimes\mathbb{H}_2}(-0.5,0.1,0.1,0)$ along i_1-axis (from 0.1 to 0.4) and i_2-axis (from 0.1 to 0.5), with a step of 0.1.

that nondegenerated $\mathcal{M}^{\mathbb{H}_2}$- and $\mathcal{J}^{\mathbb{H}_2}$-sets can be constructed. The multiquaternion algebra $\mathbb{H}_n$ can be also considered for the construction of $\mathcal{M}^{\mathbb{H}_n}$- and $\mathcal{J}^{\mathbb{H}_n}$-sets, since it can be regarded as the Clifford algebra of a suitable nondegenerate quadratic form over the base field [74], which initiates appropriate conditions of existence of nondegenerated fractal sets.

The next group of tensor product algebras is the group of octonionic tensor products, where the first of such algebras is the 16-algebra of bioctonions $\mathbb{C} \otimes \mathbb{O}$. This algebra is a simple alternative algebra [110] that contains zero-divisors [5], however, as the previous analysis showed [62], it is appropriate for constructing bioctonionic $\mathcal{M}$- and $\mathcal{J}$-sets. Other two tensor product algebras, which can be constructed over octonions, are the 32-algebra quateroctonions $\mathbb{H} \otimes \mathbb{O}$ and the 64-algebra of octooctonions $\mathbb{O}_2$. These algebras are not even alternative [110], thus, we cannot obtain nondegenerated quateroctonionic and octooctonionic $\mathcal{M}$- and $\mathcal{J}$-sets. This is also applicable for the multioctonions $\mathbb{O}_n$.

Of course, many other tensor product algebras can be constructed from split algebras obtained from complex numbers, quaternions, and octonions. However, as it was shown in several examples on split algebra derived from complex numbers (see Section 2.4 for details), due to their properties, these algebras cannot create conducive conditions for obtaining nondegenerated $\mathcal{M}$- and $\mathcal{J}$-sets. Therefore, considering also the fractal sets of $\mathcal{M}$- and $\mathcal{J}$-type based on the Cliffordean algebras (which were previously discussed in Section 2.3), our journey through the world of hypercomplex fractals has come to end.

Bibliography

[1] O. Aam. Principia cybernetica mailing-list archive: Liquid fractals, 1995. http://cfpm.org/~bruce/prncyb-l/0279.html.

[2] A. A. Albert. Tensor products of quaternion algebras. *Proceedings of the American Mathematical Society*, 35(1):65–66, 1972.

[3] D. S. Alexander, F. Iavernaro, and A. Rosa. *Early Days in Complex Dynamics: A History of Complex Dynamics in One Variable During 1906–1942*, volume 38 of *History of Mathematics*. American Mathematical Society, Providence, 2012.

[4] M. Bader. *Space-Filling Curves: An Introduction With Applications in Scientific Computing*. Springer Science & Business Media, Berlin Heidelberg, 2012.

[5] J. C. Baez. The octonions. *Bulletin of the American Mathematical Society*, 39(2):145–205, 2002.

[6] J. Barrallo. Expanding the Mandelbrot set into higher dimensions. In G. W. Hart and R. Sarhangi, editors, *Bridges 2010: Mathematics, Music, Art, Architecture, Culture*, pages 247–254, Pécs, 2010.

[7] J. Barrallo and D. M. Jones. Coloring algorithms for dynamical systems in the complex plane. In *ISAMA 99 Proceedings*, pages 31–38, San Sebastián, 1999. The University of the Basque Country.

[8] J. Barrallo and S. Sanchez. Fractals and multi layer coloring algorithms. *Visual Mathematics*, 3(10):0–0, 2001.

[9] T. Bedford, F. A. M., and M. Urbański. The scenery flow for hyperbolic Julia sets. *Proceedings of the London Mathematical Society*, 85(2):467–492, 2002.

[10] V. Berinde. *Iterative Approximation of Fixed Points*, volume 1912 of *Lecture Notes in Mathematics*. Springer, Berlin Heidelberg, 2nd edition, 2007.

[11] P. Blanchard. Complex analytic dynamics on the Riemann sphere. *Bulletin of the American Mathematical Society*, 11(1):85–141, 1984.

[12] A. A. Bogush, A. Z. Gazizov, Y. A. Kurochkin, and V. T. Stosui. Symmetry properties of quaternionic and biquaterionic analogs of Julia sets. *Ukrainian Journal of Physics*, 48(4):295–299, 2003.

[13] S. H. Boyd and M. J. Schultz. Geometric limits of Mandelbrot and Julia sets under degree growth. *International Journal of Bifurcation and Chaos*, 22(12):1250301, 2012.

[14] P. W. Carlson. Two artistic orbit trap rendering methods for newton M-set fractals. *Computers & Graphics*, 23(6):925–931, 1999.

[15] M. Carter. Visualization of the Cayley-Dickson hypercomplex numbers up to the chingons (64D), 2011. http://www.mapleprimes.com/posts/124913-Visualization-Of-The-CayleyDickson.

[16] A. Cayley. On Jacobis elliptic functions, in reply to the Rev. B. Bronwin; and on quaternions. *Philosophical Magazine*, 26(172):208–211, 1845.

[17] M. Cederwall and C. R. Preitschopf. s^7 and $\hat{S}^7$. *Communications in Mathematical Physics*, 167(2):373–393, 1995.

[18] F. Chaitin-Chatelin. The computing power of geometry. In G. A. Watson and D. F. Griffiths, editors, *Numerical Analysis 1999*, Chapman & Hall/CRC Research Notes in Mathematics Series, pages 83–92. CRC Press LLC, Boca Raton, FL, 2000.

[19] F. Chaitin-Chatelin and T. Meškauskas. Computation with hypercomplex numbers. *Nonlinear Analysis*, 47(5):3391–3400, 2001.

[20] Y. S. Chauhan, R. Rana, and A. Negi. New Julia sets of Ishikawa iterates. *International Journal of Computer Applications*, 7(13):34–42, 2010.

[21] J. Cheng and J. Tan. Generalization of 3D Mandelbrot and Julia sets. *Journal of Zhejiang University SCIENCE A*, 8(1):134–141, 2007.

[22] W. K. Clifford. Preliminary sketch on biquaternions. *Proceedings of the London Mathematical Society*, s1–4(1):381–395, 1871.

[23] W. K. Clifford. Applications of Grassmann's extensive algebra. *American Journal of Mathematics*, 1(4):350–358, 1878.

[24] J. Cockle. On certain functions resembling quaternions and on a new imaginary in algebra. *London-Dublin-Edinburgh Philosophical Magazine*, 3(33):43–59, 1848.

[25] J. Cockle. On a new imaginary in algebra. *London-Dublin-Edinburgh Philosophical Magazine*, 3(34):37–47, 1849.

[26] J. Cockle. On the symbols of algebra and on the theory of tessarines. *London-Dublin-Edinburgh Philosophical Magazine*, 3(34):406–410, 1849.

[27] J. Cockle. On impossible equations, on impossible quantities and on tessarines. *London-Dublin-Edinburgh Philosophical Magazine*, 3(37):281–283, 1850.

[28] J. H. Conway and D. A. Smith. *On Quaternions and Octonions: Their Geometry, Arithmetic, and Symmetry*. A K Peters, Wellesley, MA, 2003.

[29] W. D. Crowe, R. Hasson, P. Rippon, and P. E. D. Strain-Clark. On the structure of the Mandelbar set. *Nonlinearity*, 2(4):541–553, 1989.

[30] Y. Dang, L. H. Kauffman, and D. J. Sandin. *Hypercomplex Iterations. Distance Estimation and Higher Dimensional Fractals*, volume 17 of *Series on Knots and Everything*. World Scientific, Singapore, 2002.

[31] S. Datta. Infinite sequences in the constructive geometry of tenth-century Hindu temple superstructures. *Nexus Network Journal*, 12(3):471–483, 2010.

[32] R. L. Devaney. The fractal geometry of the Mandelbrot set 2: How to count and how to add. *Fractals*, 3(4):629–640, 1995.

[33] R. L. Devaney and F. Tangerman. Dynamics of entire functions near the essential singularity. *Ergodic Theory and Dynamical Systems*, 6(4):489–503, 1986.

[34] S. V. Dhurandhar, V. C. Bhavsar, and U. G. Gujar. Analysis of z-plane fractal images from $z \leftarrow z^{\alpha} + c$ for $\alpha < 0$. *Computers & Graphics*, 17(1):89–94, 1993.

[35] S. L. Dixon, K. L. Steele, and R. P. Burton. Generation and graphical analysis of Mandelbrot and Julia sets in more than four dimensions. *Computers & Graphics*, 20(3):451–456, 1996.

[36] R. Eglash. *African Fractals: Modern Computing and Indigenous Design*. Rutgers University Press, New Brunswick, NJ, 1999.

[37] N. Fagella. Dynamics of the complex standard family. *Journal of Mathematical Analysis and Applications*, 229(1):1–31, 1999.

[38] H. Freudenthal. Oktaven, Ausnahmegruppen und Oktavengeometrie. *Geometriae Dedicata*, 19(1):7–63, 1985.

[39] V. Garant-Pelletier and D. Rochon. On a generalized Fatou-Julia theorem in multicomplex spaces. *Fractals*, 17(3):241–255, 2009.

[40] K. Gdawiec, W. Kotarski, and A. Lisowska. Biomophs via modified iterations. *Journal of Nonlinear Science and Applications*, 9:2305–2315, 2016.

[41] T. W. Gintz. Artist's statement CQUATS — a non-distributive quad algebra for 3D renderings of Mandelbrot and Julia sets. *Computers & Graphics*, 26(2):367–370, 2002.

[42] E. F. Glynn. The evolution of the Gingerbread Man. *Computers & Graphics*, 15(4):579–582, 1991.

[43] J. Gomatham, J. Doyle, B. Steves, and M. I. Generalization of the Mandelbrot set: Quaternionic quadratic maps. *Chaos Solitons and Fractals*, 5(6):971–986, 1995.

[44] C. J. Griffin and G. C. Joshi. Octonionic Julia sets. *Chaos Solitons and Fractals*, 2(1):11–24, 1992.

[45] C. J. Griffin and G. C. Joshi. Associators in generalized octonionic maps. *Chaos Solitons and Fractals*, 3(3):307–319, 1993.

[46] C. J. Griffin and G. C. Joshi. Transition points in octonionic Julia sets. *Chaos Solitons and Fractals*, 3(1):67–88, 1993.

[47] U. G. Gujar and V. C. Bhavsar. Fractals from $z \leftarrow z^\alpha + c$ in the complex c-plane. *Computers & Graphics*, 15(3):441–449, 1991.

[48] U. G. Gujar, V. C. Bhavsar, and N. Vangala. Fractals from $z \leftarrow z^\alpha + c$ in the complex z-plane. *Computers & Graphics*, 16(1):45–49, 1992.

[49] S. Halayka. Some visually interesting non-standard quaternion fractal sets. *Chaos Solitons and Fractals*, 41(5):2842–2846, 2009.

[50] A. J. Hanson. *Visualizing Quaternions*. The Morgan Kaufmann Series in Interactive 3D Technology. Morgan Kaufmann Publishers, San Francisco, CA, 2006.

[51] J. C. Hart, L. H. Kauffman, and D. J. Sandin. Interactive visualization of quaternion Julia sets. In *Proceedings of the First IEEE Conference on Visualization*, pages 209–218, San Francisco, CA, 1990.

[52] J. C. Hart, G. W. Lescinsky, D. J. Sandin, T. A. DeFanti, and L. H. Kauffman. Scientific and artistic investigation of multi-dimensional fractals on the AT & T pixel machine. *The Visual Computer*, 9(7):346–355, 1993.

[53] J. C. Hart, D. J. Sandin, and L. H. Kauffman. Ray tracing deterministic 3-D fractals. *Computer Graphics*, 23(3):289–296, 1989.

[54] S.-M. Heinemann and B. O. Stratmann. Geometric exponents for hyperbolic Julia sets. *Illinois Journal of Mathematics*, 45(3):775–785, 2001.

[55] Z. Huang and J. Wang. On the radial distribution of Julia sets of entire solutions of $f^{(n)} + a(z)f = 0$. *Journal of Mathematical Analysis and Applications*, 387(2):1106–1113, 2012.

[56] S. Ishikawa. Fixed points by a new iteration method. *Proceedings of the American Mathematical Society*, 44:147–150, 1974.

[57] V. G. Ivancevic and T. T. Ivancevic. *Complex Nonlinearity: Chaos, Phase Transitions, Topology Change and Path Integrals*. Springer, Berlin, 2008.

[58] M. Jafari. Split semi-quaternions algebra in semi-Euclidean 4-space. *Cumhuriyet Science Journal*, 36(1):70–77, 2015.

[59] M. Jafari and Y. Yayli. Generalized quaternions and their algebraic properties. *Communications de la Faculté des Sciences de l'Université d'Ankara. Séries A1. Mathematics and Statistics*, 64(1):15–27, 2015.

[60] S. M. Kang, W. Nazeer, M. Tanveer, and A. A. Shahid. New fixed point results for fractal generation in Jungck Noor orbit with s-convexity. *Journal of Function Spaces*, 2015:963016, 2015.

[61] S. M. Kang, A. Rafiq, A. Latif, A. A. Shahid, and Y. C. Kwun. Tricorns and Multicorns of s-iteration scheme. *Journal of Function Spaces*, 2015:417167, 2015.

[62] A. Katunin. On the symmetry of bioctonionic Julia sets. *Journal of Applied Mathematics and Computational Mechanics*, 12(2):23–28, 2013.

[63] A. Katunin. Visualization of fractals based on regular convex polychora. *Journal for Geometry and Graphics*, 19(1):1–17, 2015.

[64] A. Katunin. *Analysis of 4D Hypercomplex generalizations of Julia sets*, volume 9972 of *Lecture Notes in Computer Science*, pages 627–635. Springer International Publishing, 2016.

[65] A. Katunin. On the convergence of multicomplex M–J sets to the Steinmetz hypersolids. *Journal of Applied Mathematics and Computational Mechanics*, 15(3):67–74, 2016.

[66] A. Katunin and K. Fedio. On a visualization of the convergence of the boundary of generalized Mandelbrot set to $(n-1)$-sphere. *Journal of Applied Mathematics and Computational Mechanics*, 14(1):63–69, 2015.

[67] M. A. Krasnosel'ski. Two observations about the method of succesive approximations. *Uspekhi Matematicheskikh Nauk*, 10:123–127, 1955.

[68] V. V. Kravchenko. *Applied Quaternionic Analysis*, volume 28 of *Research and Exposition in Mathematics*. Heldermann Verlag, Lemgo, 2003.

[69] A. Kricker and G. C. Joshi. Bifurcation phenomena of the non-associative octonionic quadratic. *Chaos Solitons and Fractals*, 5(5):761–782, 1995.

[70] J. Kudrewicz. *Fractals and Chaos*. Wydawnictwa Naukowo-Techniczne, Warsaw, 4th edition, 2007. [in Polish].

[71] Y. A. Kurochkin and S. Y. Zhukovich. Set symmetry, generated by octonion analog of Julia–Fatou algorithm. *Vestnik Brestskaga Universiteta — Serya 4. Fizika Matematyka*, 2:74–49, 2010. [in Russian].

[72] M. Lakner and P. Petek. The one-equator property. *Experimental Mathematics*, 6(2):109–115, 1997.

[73] M. Levin. Morphogenetic fields in embryogenesis, regeneration, and cancer: non-local control of complex patterning. *Biosystems*, 109(3):243–261, 2012.

[74] D. W. Lewis. A note on Clifford algebras and central division algebras with involution. *Glasgow Mathematical Journal*, 26(2):171–176, 1985.

[75] J. Leys. Biomorphic art: An artist's statement. *Computers & Graphics*, 26:977–979, 2002.

[76] I.-H. Lin. *Classical Complex Analysis: A Geometric Approach*, volume 2. World Scientific, Singapore, 2011.

[77] S. Liu, X. Cheng, C. Lan, W. Fu, J. Zhou, Q. Li, and G. Gao. Fractal property of generalized M-set with rational number exponent. *Applied Mathematics and Computation*, 220:668–675, 2013.

[78] M. E. Luna-Elizarrarás, M. Shapiro, D. C. Struppa, and A. Vajiac. Bicomplex numbers and their elementary functions. *CUBO A Mathematical Journal*, 14(2):61–80, 2012.

[79] B. B. Mandelbrot. *The Fractal Geometry of Nature*. W. H. Freeman and Co., New York, NY, 1982.

[80] W. R. Mann. Mean value methods in iteration. *Proceedings of the American Mathematical Society*, 4:506–510, 1953.

[81] E. Martineau and D. Rochon. On a bicomplex distance estimation for the Tetrabrot. *International Journal of Bifurcation and Chaos*, 15(9):3039–3050, 2005.

[82] C. Matteau and D. Rochon. The inverse iteration method for Julia sets in the 3-dimensional space. *Chaos Solitons and Fractals*, 75:272–280, 2015.

[83] C. McMullen. Area and Hausdorff dimension of Julia sets of entire functions. *Transactions of the American Mathematical Society*, 300(1):329–342, 1987.

[84] W. Metzler. The 'mystery' of the quadratic Mandelbrot set. *American Journal of Physics*, 62(9):813–814, 1994.

[85] N. S. Mojica, J. Navarro, P. C. Marijuán, and R. Lahoz-Beltra. Cellular 'bauplants': Evolving unicellular forms by means of Julia sets and Pickover biomorphs. *Biosystems*, 98:19–30, 2009.

[86] M. Moore. Symmetrical intersections of right circular cylinders. *The Mathematical Gazette*, 58(405):181–185, 1974.

[87] H. Mortazaasl and M. Jafari. A study on semi-quaternions algebra in semi-Euclidean 4-space. *Mathematical Sciences and Applications E-Notes*, 1(2):20–27, 2013.

[88] K. Nagashima and H. Morimatsu. 3D representation of the Mandelbrot set. *The Visual Computer*, 10(6):356–359, 1994.

[89] S. Nakane. Connectedness of the tricorn. *Ergodic Theory and Dynamical Systems*, 13(2):349–356, 1993.

[90] S. Nakane and D. Schleicher. On Multicorns and Unicorns I: Antiholomorphic dynamics, hyperbolic components and real cubic polynomials. *International Journal of Bifurcation and Chaos*, 13(10):2825–2844, 2003.

[91] W. Nazeer, S. M. Kang, M. Tanveer, and A. A. Shahid. Fixed point results in the generation of Julia and Mandelbrot sets. *Journal of Inequalities and Applications*, 2015:298, 2015.

[92] A. Negi, M. Rani, and R. Chugh. Julia and Mandelbrot sets in Noor orbit. *Applied Mathematics and Computation*, 228:615–631, 2014.

[93] A. Negi, M. Rani, and P. K. Mahanti. Computer simulation of the behaviour of Julia sets using switching processes. *Chaos Solitons and Fractals*, 37(4):1187–1192, 2008.

[94] A. V. Norton. Generation and display of geometric fractals in 3-D. *Computer Graphics*, 16(3):61–67, 1982.

[95] A. V. Norton. Julia sets in the quaternions. *Computers & Graphics*, 13(2):267–278, 1989.

[96] T. V. Papathomas and B. Julesz. Animation with fractals from variations on the Mandelbrot set. *The Visual Computer*, 3(1):23–26, 1987.

[97] P.-O. Parisé and D. Rochon. A study on dynamics of the tricomplex polynomial $\eta^p + c$. *Nonlinear Dynamics*, 82(1):157–171, 2015.

[98] H.-O. Peitgen, H. Jurgens, and D. Saupe. *Chaos and Fractals: New Frontiers of Science*. Springer Science & Business Media, New York, NY, 2013.

[99] P. Petek. *Chaotic Dynamics. Theory and Practice*, volume 298 of *NATO ASI Series*, chapter On the quaternionic Julia sets, pages 53–58. Springer Science & Business Media, New York, NY, 1992.

[100] P. Petek. Circles and periodic points in quaternionic Julia sets. *Open Systems & Information Dynamics*, 4(4):487–492, 1997.

[101] C. A. Pickover. Biomorphs: Computer displays of biological forms generated from mathematical feedback loops. *Computer Graphics Forum*, 5:313–316, 1986.

[102] C. A. Pickover, editor. *The Pattern Book: Fractals, Art, and Nature*. World Scientific, Singapore, 1995.

[103] G. B. Price. *An Introduction to Multicomplex Spaces and Functions*, volume 140 of *Pure and Applied Mathematics*. Marcel Dekker Inc., New York, NY, 1991.

[104] M. Rani and V. Kumar. Superior Julia set. *Journal of the Korea Society of Mathematical Education Series D: Research in Mathematical Education*, 8(4):261–277, 2004.

[105] M. Rani and A. Negi. New Julia sets for complex Carotid-Kundalini function. *Chaos Solitons and Fractals*, 36(2):226–236, 2008.

[106] I. M. Rian, J.-H. Park, H. U. Ahn, and D. Chang. Fractal geometry as the synthesis of Hindu cosmology in Kandariya Mahadev temple, Khajuraho. *Building and Environment*, 42(12):4093–4107, 2007.

[107] D. Rochon. A generalized Mandelbrot set for bicomplex numbers. *Fractals*, 8(4):355–368, 2000.

[108] D. Rochon. On a generalized Fatou-Julia theorem. *Fractals*, 11(3):213–219, 2003.

[109] A. Rosa. Methods and applications to display quaternion Julia sets. *Differential Equations and Control Processes*, 4:1–22, 2005.

[110] B. Rosenfeld. *Geometry of Lie Groups*, volume 393 of *Mathematics and its applications*. Springer Science & Business Media, Dordrecht, Holland, 1997.

[111] G. Rottenfusser and D. Schleicher. Escaping points of the cosine family. In P. J. Rippon and G. M. Stallard, editors, *Transcendental Dynamics and Complex Analysis*, volume 348 of *London Mathematical Society Lecture Note Series*, pages 396–424. Cambridge University Press, Cambridge, MA, 2008.

[112] L. Sabinin, L. Sbitneva, and I. Shestakov, editors. *Non-Associative Algebra and Its Applications*, volume 246 of *Lecture Notes in Pure and Applied Mathematics*. CRC Press, Boca Raton, FL, 2006.

[113] D. Schleicher and J. Zimmer. Escaping points of exponential maps. *Journal of the London Mathematical Society*, 67(2):380–400, 2003.

[114] P. Senn. The Mandelbrot set for binary numbers. *American Journal of Physics*, 58(10):1018, 1990.

[115] K. W. Shirriff. An investigation of fractals generated by $z \leftarrow 1/z^n + c$. *Computers & Graphics*, 17(5):603–607, 1993.

[116] Y.-Y. Sun, P. Li, and Z.-X. Lu. Generalized quaternion M sets and Julia sets perturbed by dynamical noises. *Nonlinear Dynamics*, 82(1):143–156, 2015.

[117] K. Trivedi. Hindu temples: Models of a fractal universe. *The Visual Computer*, 5(4):243–258, 1989.

[118] X. Wang and T. Jin. Hyperdimensional generalized M–J sets in hypercomplex number space. *Nonlinear Dynamics*, 73(1):843–852, 2013.

[119] X.-Y. Wang and W. J. Song. The generalized M-J sets for bicomplex numbers. *Nonlinear Dynamics*, 72(1):17–26, 2013.

[120] X.-Y. Wang and Y.-Y. Sun. The general quaternionic M–J sets on the mapping $z \leftarrow z^\alpha + c$ ($\alpha \in \mathbf{N}$). *Computers & Mathematics with Applications*, 53(11):1718–1732, 2007.

[121] R. Ye. Another choice for orbit traps to generate artistic fractal images. *Computers & Graphics*, 26(4):629–633, 2002.

[122] H. Zeitler. Iterations over quaternions. *Mathematica Pannonica*, 15(1):85–103, 2004.

[123] A. Zireh. A generalized Mandelbrot set of polynomials of type e_d for bicomplex numbers. *Georgian Mathematical Journal*, 15(1):189–194, 2008.

Index